# Feature Papers in Extractive Metallurgy

# Feature Papers in Extractive Metallurgy

Guest Editor

**Srecko Stopic**

MDPI

Basel • Beijing • Wuhan • Barcelona • Belgrade • Novi Sad • Cluj • Manchester

*Guest Editor*
Srecko Stopic
IME Process Metallurgy and
Metal Recycling Department
RWTH Aachen University
Aachen
Germany

*Editorial Office*
MDPI AG
Grosspeteranlage 5
4052 Basel, Switzerland

This is a reprint of the Special Issue, published open access by the journal *Metals* (ISSN 2075-4701), freely accessible at: https://www.mdpi.com/journal/metals/special_issues/YBPM42EBT9.

For citation purposes, cite each article independently as indicated on the article page online and as indicated below:

| Lastname, A.A.; Lastname, B.B. Article Title. *Journal Name* **Year**, *Volume Number*, Page Range. |
|---|

**ISBN 978-3-7258-4681-8 (Hbk)**
**ISBN 978-3-7258-4682-5 (PDF)**
**https://doi.org/10.3390/books978-3-7258-4682-5**

# Contents

## About the Editor

**Srecko Stopic**

Srecko Stopic was born on April 3, 1965 in Uzice, Serbia, where he completed primary and secondary school. He graduated in 1991, and obtained a master's degree in 1994 and a doctorate in 1997 at the Faculty of Technology and Metallurgy in Belgrade, Serbia, majoring in non-ferrous metallurgy. In 2001, he received the Alexander von Humboldt research scholarship for Scientific Excellence and transferred to the RWTH Aachen University in Germany, where he leads projects in the field of extractive metallurgy at the Institute for Process Metallurgy and Metal Recycling (EURO-TITAN). He runs courses in extractive metallurgy of rare earth elements and Hydrometallurgy and electrolysis at the RWTH Aachen University.

metals MDPI

*Editorial*

# Feature Papers in Extractive Metallurgy

**Srecko Stopic**

IME Process Metallurgy and Metal Recycling, RWTH Aachen University, Intzestrasse 3, 52056 Aachen, Germany; sstopic@ime-aachen.de; Tel.: +49-17678261674

## 1. Introduction

Unit metallurgical operation processes are usually separated into three categories: (1) hydrometallurgy (leaching, mixing, neutralization, precipitation, cementation, and crystallization) [1], (2) pyrometallurgy (roasting and smelting) [2], and (3) electrometallurgy (aqueous electrolysis and molten salt electrolysis) [3]. In hydrometallurgy, the target metal is first transferred from ores and concentrates to a solution using selective dissolution (leaching; dry digestion) under an atmospheric pressure below 100 °C, a high pressure (40–50 bar), and high temperature (below 270 °C) in an autoclave and tube reactor. The purification of the obtained solution is performed using neutralization agents such as sodium hydroxide and calcium carbonate, or more selective precipitation agents such as sodium carbonate and oxalic acid. The separation of metals is possible using liquids/liquid processes (solvent extraction in a mixer-settler) and solid–liquid (filtration in filter-press under high pressure) methods. Crystallization is the process by which a metallic compound is converted from a liquid into a solid crystalline state via a supersaturated solution. The final step is metal production using electrochemical methods (aqueous electrolysis for basic metals such as copper, zinc, and silver and molten salt electrolysis for rare earth elements and aluminum). Advanced processes for metal production, such as ultrasonic spray pyrolysis and microwave-assisted leaching, can be combined with reduction processes.

## 2. Results and Discussion

China's crude steel production has reached 1.019 billion tons, and the energy consumption of China's steel industry has reached 561 million tons of coal. China's steel industry, with its dominant reliance on coal for energy and the primary use of blast furnaces and converters in production processes, as well as its massive output, has become the main field for achieving China's "carbon peaking" and "carbon neutrality" goals. Firstly, this article summarizes the current production status of the steel industry and its situation regarding carbon emissions. Secondly, it discusses the dual carbon policies based on the national and steel industry levels and outlines the future directions for China's steel industry [Contribution 1]. Subsequently, it analyzes the current state of research and application of mature and emerging low-carbon technologies in the industry and details the low-carbon plans of China's steel companies using the low-carbon technology roadmaps of two representative steel companies as examples. Finally, the article gives policy suggestions for further carbon reduction in the industry. The purpose of this paper is to show the efforts and contributions of China's steel industry to the early realization of its "carbon peaking" and "carbon neutrality" goals.

There is a long process for transforming bauxite, a critical raw material, into a substance with the required properties of cast aluminum alloys for use in electro automotive parts. Thanks to its unique properties, aluminum has become the material of choice for clean technology manufacturers in applications such as use in the automotive industry, renewable energy, batteries, electrical systems, resource-saving packaging, energy-efficient buildings, and clean mobility [4]. The restructuring of the economy, the oil crisis, air pollution, and global warming are some of the factors that have moved the automotive industry towards electrification since the beginning of the 21st century. This paper aims to highlight the required properties of cast aluminum alloys applied to the production of electro automotive parts, such as their mechanical and thermophysical properties, dimensional stability, corrosion resistance, electromagnetic compatibility, and crashworthiness. Furthermore, this paper discusses which of the cast aluminum–silicon alloys, as well as heat treatments and casting processes, are most suitable.

This paper considers the impact of copper mining-influenced water and metallurgical wastewater on the surface water in the Bor area, Serbia. Sampling, realized through four campaigns (2020–2021), confirmed that both types of water, discharged without appropriate treatment into the Bor River, had a signific impact on the concentration of metal ions, pH, and electrical conductivity on the watercourse in the Bor area [Contribution 2]. The highest concentrations of the following metal ions, Cu—271 mg/L, As—25,991 μg/L, Ni—13,856 μg/L, Cd—2627 μg/L, and Pb—2855 μg/L, were registered in the metallurgical wastewater samples. After changes occurred in the copper production process by stopping the discharge of untreated wastewater into the Bor River, the concentrations of monitored elements drastically decreased. In the period 2022–2024, the concentration values for Cu, As, and Pb ions were below the maximum allowable value, and the concentration values of Ni and Cd ions had also decreased. The values for pH and electrical conductivity were in the maximum allowable range. The return of wastewater to the copper production process would lead to both a reduction in the primary water consumption and a reduction in the negative impact on the environment [5].

In modern times, more and more different materials (including alloys) are in direct contact with human electrolytes (sweat, saliva, lymph, blood, etc.). One of the most important properties for the use of these materials is therefore their chemical inertness or resistance to corrosion when they are in contact with human electrolytes. Consequently, during the development of such new materials, it is necessary to study and understand their basic electrochemical behaviour in a given environment [Contribution 3]. The purpose of this research was to monitor the electrochemical behaviour of the new Au-Ge alloy in artificial sweat and saliva solutions, depending on the electrolyte composition and exposure time. This new alloy represents a potential material for use in dentistry or for jewellery [6,7]. The obtained results of the study show that the immersion time and the pH value have a significantly greater influence on the corrosion resistance of the new Au-Ge alloy than on the composition of the electrolyte solution. The results of the SEM/EDX analysis additionally confirm the main results of the electrochemical measurements.

Copper smelting slag is a significant potential resource for cobalt and copper. The recovery of copper and cobalt from copper slag could significantly augment the supply of these metals, which are essential in facilitating the transition to green energy while simultaneously addressing environmental concerns regarding slag disposal [Contribution 4]. However, the complex mineral composition of copper slag poses an enormous challenge. This study investigated the mineralogical and chemical characteristics of copper slag, which are vital for devising the most effective processing techniques. XRD and FESEM-EDS were employed to examine the morphologies of copper slag before and after the reduction process. The effects of borax and charcoal (carbocatalytic) reduction on phase transformation were evaluated. The XRD analysis revealed that the primary phases in the copper slag were $Fe_2SiO_4$ and $Fe_3O_4$. The FESEM-EDS analysis verified the presence of these phases and yielded supplementary details regarding metal embedment in the $Fe_2SiO_4$, $Fe_3O_4$, and

Cu phases [8]. The carbocatalytic reduction process expedited the transformation of copper slag microstructures from crystalline dendritic into amorphous and metallic phases. Finally, leaching experiments demonstrated the potential benefits of carbocatalytic reduction by yielding high extractions of Cu, Co, and Fe.

In the pre-reduction cyclone of the HIsarna process, both thermal decomposition and gas reduction in the injected iron ores occur simultaneously at gas temperatures of 1723–1773 K. In this study, the kinetics of the thermal decomposition of three iron ores (namely OreA, OreB, and OreC) for HIsarna ironmaking were analyzed as an isolated process with a symmetrical thermogravimetric analyser (TGA) under an inert atmosphere [Contribution 5]. Using various methods, the chemical and mineralogical composition, particle size distribution, morphology, and phase distribution of the ores were analyzed [Contribution 6]. The ores differ in their mineralogy and morphology, where OreA only contains hematite as the iron-bearing phase and OreB and OreC include goethite and hematite. To obtain the kinetic parameters in non-isothermal conditions, the Coats–Redfern Integral Method was applied for heating rates of 1, 2, and 5 K/min and a maximum temperature of 1773 K. The TGA results indicate that goethite and hematite decomposition occur as a two-stage process in an inert atmosphere of Ar. The proposed reaction mechanism for the first stage of goethite decomposition is chemical reaction with an activation energy ranging from 46.55 to 60.38 kJ/mol for OreB and from 69.90 to 134.47 kJ/mol for OreC. The proposed reaction mechanism for the second stage of goethite decomposition is diffusion, showing an activation energy ranging between 24.43 and 44.76 kJ/mol for OreB and between 3.32 and 23.29 kJ/mol for OreC. In terms of hematite decomposition, only the first stage was analyzed. The proposed reaction mechanism is chemical reaction control. OreA shows an activation energy of 545.47 to 670.50 kJ/mol, OreB of 587.68 to 831.54 kJ/mol, and OreC of 424.31 to 592.32 kJ/mol.

The electrochemical performances of an oxygen-evolving anode produced by the reactivation of waste Ti substrate by a typical $IrO_2$-$Ta_2O_5$ coating are correlated to the textural (non)uniformities of the coating and its exhaustion state [Contribution 7]. Coating degradation is considered operational loss of the activity in a metal electrowinning process. It was found that (pseudo)capacitive performances can vary over the coating surface by 20–30% and depend on the type of dynamics of the input perturbation: constant through cyclic voltammetry (CV) or discontinuous time-dependent through electrochemical impedance spectroscopy (EIS). CV-EIS data correlation enabled profiling of the capacitive properties through the depth of a coating and over its surface. The correlation was confirmed by the findings for the analysis of coating activity for an oxygen evolution reaction, finally resulting in the reliable proposition of a mechanism for the operational loss of the anode. It was found that the less compact and thicker coating parts performed better and operated more efficiently, especially at lower operational current densities.

SmCo magnets are a common material utilized in advanced technological applications. These magnets contain elevated concentrations of Sm and Co within their structural compositions. Given that both Sm and Co are classified as critical metals, the recycling of these magnets after their operational lifespan is of significant economic and environmental importance [9]. Hydrometallurgical recycling processes represent an effective method for the recycling of these magnets. In this study [Contribution 8], a pH-controlled selective precipitation method was developed using two $HNO_3$ solutions with distinct oxidizing properties for the recovery of Sm and Co from end-of-life (EoL) SmCo magnets. In the initial stage of the process, the magnets were leached in a low-oxidizing 2M $HNO_3$ solution, with a 1:30 ($w/v$) solid-to-liquid ratio at 20 °C. This step was undertaken to ensure the dissolution of Fe, thereby creating an environment conducive to its removal from the solution. The leaching experiments resulted in dissolution efficiencies of 95%, 96%, and 96% for

Sm, Co, and Fe, respectively. In the second stage, a leaching experiment was performed using 3M $HNO_3$ with a 1:10 solid-to-liquid ratio at 60 °C. Under these conditions, Sm and Co achieved dissolution efficiencies of 99%, while Fe remained undissolved in the solid phase due to hydrolysis at the high temperature, thus increasing the solution purity. In the precipitation process, the pH of both leachates was initially adjusted to 4 to precipitate impurities such as Fe ions. As a result of precipitation at pH 4 in the 2M $HNO_3$ leachate, Fe ions were almost completely removed. This was followed by selective Sm precipitation in the pH range of 5–6.5 using $NH_4HCO_3$. The highest purity of Sm precipitation was achieved when the pH reached 6.5. An increase in Sm precipitation efficiency was observed with increasing pH, with an efficiency of 12.75% at pH 5, which rose to 82.37% at pH 6.5. Furthermore, although the precipitation efficiency of Co increased from 6.25% to 10% within this pH range, no significant difference in the extent of this increase was observed. In the case of the 3M $HNO_3$ leachate, the Sm precipitation efficiency at pH 5 was 44.28%, while at pH 6.5, nearly all of the Sm ions were precipitated. The co-precipitation efficiency at pH 5 was 1.89%, increasing to 36.43% at pH 6.5. This increase in co-precipitation was attributed to the system's Eh value, which reflects the enhanced oxidizing properties of the 3M $HNO_3$ leach solution. The results of the study indicate that as the oxidizing strength of the solution increased, the co-precipitation rate also increased with rising pH.

The efficient extraction of zinc from polymetallic concentrates is crucial for the metallurgical industry. Traditional leaching techniques often rely on strong oxidizing agents, which can be wasteful and environmentally harmful [10]. While cyclic oxidation systems like the $Fe^{3+}/Fe^{2+}$ pair are known, they often fail to achieve high leaching rates, especially when the raw material contains multiple sulphide minerals. In this study, we developed a novel oxidation system using manganese dioxide ($MnO_2$) as the primary oxidizing agent and potassium iodide (KI) as a supporting material to create an $I_2/I^-$ oxidation cycle in a sulphuric acid medium, at an atmospheric pressure between 40 °C and 80 °C. Leaching experiments were conducted under varying temperatures and KI doses [Contribution 9]. The results demonstrated that for the $MnO_2$-KI system, a zinc leaching degree of 89.78% was achieved after 3 h of leaching at 80 °C, and kinetic studies indicated that the leaching process is diffusion controlled (through the thin film), with an activation energy of 27.65 kJ $mol^{-1}$. Moreover, this system offers an improved method for separating iodine from the leachate upon completion, enhancing the overall process efficiency. It also opens up opportunities to test other primary oxidizing agents in combination with iodide salts. These findings suggest that the $MnO_2$-KI oxidation system offers a promising approach for improving zinc recovery from sphalerite concentrates.

With the growing demand for metals driven by technological advancements and population growth, recycling lithium-ion batteries has become vital for protecting the environment and recovering valuable materials [11]. Developing sustainable recycling technologies is now more essential than ever [12]. This paper focuses on using electrodialysis to process a leach solution of $LiNi_{0.33}Mn_{0.33}Co_{0.33}O_2$ (NMC 111) cathode materials leached with citric acid. This study demonstrates that the complexing properties of citrate anions contribute to the efficient separation of Li from Ni, Co, and Mn via electrodialysis [11]. This is achieved by promoting the formation of anionic species for Ni, Co, and Mn while maintaining Li in its cationic form. The leach solution was produced under the following optimal experimental conditions to reach a final pH of 5 and high leaching efficiency: a citric acid concentration of 1 mol $L^{-1}$, a leaching temperature of 45 °C, a leaching time of 5 h, a liquid/solid ratio of 100 g/L, and 8 vol.% $H_2O_2$. These conditions resulted in leaching efficiencies of 89.3% for Ni, 95.1% for Co, 77.1% for Mn, and 92.9% for Li. This solution led to the formation of a lithium-rich supernatant and a precipitate [Contribution 10]. The supernatant was then used as the feed solution for electrodialysis. Pure lithium was successfully separated with a faradic efficiency of 71.4% with a commercial cation-exchange membrane. This strategy enables selective lithium recovery while minimizing membrane fouling during the process.

## 3. Conclusions

Unit metallurgical operation processes are usually separated into three categories and are explained in detail in this paper. Advances in the understanding of different technologies in the extractive metallurgy of non-ferrous metals and iron and steel were studied in this work.

**Conflicts of Interest:** The author declares no conflicts of interest.

**List of Contributions:**

1. Qiao, Y.; Wang, G. Recent Status of Production, Administration Policies, and Low-Carbon Technology Development of China's Steel Industry. *Metals* **2024**, *14*, 480. https://doi.org/10.3390/met14040480.
2. Marković, R.; Marjanović, V.; Stevanović, Z.; Gardić, V.; Petrović, J.; Kovačević, R.; Štirbanović, Z.; Friedrich, B. Importance of Changes in the Copper Production Process through Mining and Metallurgical Activities on the Surface Water Quality in the Bor Area, Serbia. *Metals* **2024**, *14*, 649. https://doi.org/10.3390/met14060649.
3. Vastag, G.; Majerič, P.; Lazić, V.; Rudolf, R. Electrochemical Behaviour of an Au-Ge Alloy in an Artificial Saliva and Sweat Solution. *Metals* **2024**, *14*, 668. https://doi.org/10.3390/met14060668.
4. Chanda Phiri, T.; Singh, P.; Nikoloski, A. Mineralogical Characterisation of Copper Slag and Phase Transformation after Carbocatalytic Reduction for Hydrometallurgical Extraction of Copper and Cobalt. *Metals* **2024**, *14*, 1119. https://doi.org/10.3390/met14101119.
5. Leerhoff, P.; Brouwer, J.; Armaki, A.; Zeilstra, C.; Meijer Kvan der Stel, J.; Shoshan, T.; Neslihan, A.; Dogan, N.; Yang, Y. Characterisation of Varying Iron Ores and Their Thermal Decomposition Kinetics Under HIsarna Ironmaking Conditions. *Metals* **2024**, *14*, 1271. https://doi.org/10.3390/met14111271.
6. Raabe, D. The Materials Science behind Sustainable Metals and Alloys. *Chem. Rev.* **2023**, *123*, 2436–2608.
7. Bošnjaković, J.; Vladimir Panić, V.; Stevanović, M; Stopic, S.; Stevanović, J.; Grgur, B.; Šekularac, G. Revealing the Surface and In-Depth Operational Performances of Oxygen-Evolving Anode Coatings: A Guideline for the Synthesis of Inert Durable Anodes in Metal Electrowinning from Acid Solutions. *Metals* **2024**, *14*, 1339. https://doi.org/10.3390/met14121339.
8. Uysal, E.; Emil-Kaya, E.; Dursun, H.; Papakci, M.; Gürmen, S. Bernd Friedrich; Recovery of Samarium from Waste SmCo Magnets via Selective Precipitation with Ammonium Bicarbonate: Optimization of Process Efficiency. *Metals* **2024**, *14*, 1363. https://doi.org/10.3390/met14121363.
9. Bugarčić, M.; Jovanović, A.; Anđić, D; Jelić, I.; Miletić, M.; Marković, B.; Sokić, M. Kinetics of Sphalerite Leaching by $MnO_2$-KI Oxidation System in Sulfuric Acid. *Metals* **2025**, *15*, 50. https://doi.org/10.3390/met15010050.
10. Badre-Eddine, S.; Muhr, L.; Chagnes, A. Innovative Integration of Citric Acid Leaching and Electrodialysis for Selective Lithium Recovery from NMC Cathode Material. *Metals* **2025**, *15*, 598. https://doi.org/10.3390/met15060598.

## References

1. Cheema, H.A.; Ilyas, S.; Farhan, M.; Yang, J.-H.; Kim, H. Extraction of Fe and Al from Red Mud Using Carbothermic Reduction Followed by Water Leaching. *J. Ind. Eng. Chem.* **2024**, *136*, 201–210. [CrossRef]
2. Habibi, H.; Piruzian, D.; Shakibania, S.; Pourkarimi, Z.; Mokmeli, M. The Effect of Carbothermal Reduction on the Physical and Chemical Separation of the Red Mud Components. *Miner. Eng.* **2021**, *173*, 107216. [CrossRef]

3. Cvetkovic, V.; Jovicevic, N.; Vukicevic, N.; Jovicevic, J. Aluminium/zirconium alloys obtained by Al underpotential deposition onto Zr from low temperature $AlCl_3$+NaCl molten salts. *J. Serbian Chem. Soc.* **2019**, *84*, 1329–1344. [CrossRef]
4. Djurdjevic, M.; Manasijevic, S.; Mihailović, M.; Stopic, S. From Bauxite as a Critical Material to the Required Properties of Cast Aluminum Alloys for Use in Electro Automotive Parts. *Metals* **2023**, *13*, 1796. [CrossRef]
5. Usman, M.O.; Aturagaba, G.; Ntale, M.; Nyakairu, G.W. A review of adsorption techniques for removal of phosphates from wastewater. *Water Sci. Technol.* **2022**, *86*, 3113–3132. [CrossRef] [PubMed]
6. Majeric, P.; Rudolf, R. Advances in Ultrasonic Spray Pyrolysis Processing of Noble Metal Nanoparticles—Review. *Materials* **2020**, *13*, 3485. [CrossRef] [PubMed]
7. Tsai, S.C.; Song, Y.L.; Tsai, C.S.; Yang, C.C.; Chiu, W.Y.; Lin, H.M. Ultrasonic Spray Pyrolysis for Nanoparticles Synthesis. *J. Mater. Sci.* **2004**, *39*, 3647–3657. [CrossRef]
8. Souza Filho, I.R.; Ma, Y.; Kulse, M.; Ponge, D.; Gault, B.; Springer, H.; Raabe, D. Sustainable Steel through Hydrogen Plasma Reduction of Iron Ore: Process, Kinetics, Microstructure, Chemistry. *Acta Mater.* **2021**, *213*, 116971. [CrossRef]
9. Xu, T.; Zhang, X.; Lin, Z.; Lü, B.; Ma, C.; Gao, X. Recovery of Rare Earth and Cobalt from Co-Based Magnetic Scraps. *J. Rare Earths* **2010**, *28*, 485–488. [CrossRef]
10. Binnemans, K.; Jones, P.T.; Blanpain, B.; Van Gerven, T.; Pontikes, Y. Towards Zero-Waste Valorisation of Rare-Earth-Containing Industrial Process Residues: A Critical Review. *J. Clean. Prod.* **2015**, *99*, 17–38.
11. Villen-Guzman, M.; Arhoun, B.; Vereda-Alonso, C.; Gomez-Lahoz, C.; Rodriguez-Maroto, J.M.; Paz-Garcia, J.M. Electrodialytic processes in solid matrices. New insights into battery recycling. A review. *J. Chem. Technol. Biotechnol.* **2019**, *94*, 1727–1738. [CrossRef]
12. Gerold, E.; Schinnerl, C.; Antrekowitsch, H. Critical Evaluation of the Potential of Organic Acids for the Environmentally Friendly Recycling of Spent Lithium-Ion Batteries. *Recycling* **2022**, *7*, 4. [CrossRef]

*metals*

MDPI

*Article*

# From Bauxite as a Critical Material to the Required Properties of Cast Aluminum Alloys for Use in Electro Automotive Parts

**Mile Djurdjevic [1], Srecko Manasijevic [2], Marija Mihailović [3] and Srecko Stopic [4,*]**

1 Nemak Linz GmbH, Zeppelinstrasse 24, A-4030 Linz, Austria; mile.djurdjevic@nemak.com
2 Lola Institute, Kneza Viseslava 70a, 11000 Belgrade, Serbia; srecko.manasijevic@li.rs
3 Institute of Chemistry, Technology and Metallurgy, National Institute of the Republic of Serbia, University of Belgrade, Njegoseva 12, 11000 Belgrade, Serbia; marija.mihailovic@ihtm.bg.ac.rs
4 Institute for Process Metallurgy and Metal Recycling, RWTH Aachen University, Intzestrasse 3, 51056 Aachen, Germany
* Correspondence: sstopic@ime-aachen.de; Tel.: +49-17678261674

**Abstract:** There is a long process to transform bauxite, a critical raw material, into a substance with the required properties of cast aluminum alloys for use in electro automotive parts. Thanks to its unique properties, aluminum has become the material of choice for clean technology manufacturers in applications such as use in the automotive industry, renewable energy, batteries, electrical systems, resource-saving packaging, energy efficient buildings and clean mobility. Restructuring of the economy, the oil crisis, air pollution and global warming are some of the factors that have moved the automotive industry towards electrification since the beginning of the 21st century. This paper aims to highlight the required properties of cast aluminum alloys applied to the production of electro automotive parts, such as their mechanical and thermophysical properties, dimensional stability, corrosion resistance, electromagnetic compatibility and crashworthiness. Furthermore, this paper discusses which of the cast aluminum–silicon alloys, as well as the heat treatments and casting processes, are most suitable.

**Keywords:** electro mobility parts; casting processes; dimensional stability; corrosion resistance; electromagnetic compatibility; crashworthiness

## 1. Introduction

Native elements such as gold and silver, lead, copper, tin and mercury have been known since ancient times, but the first scientific discovery of an element occurred in the 17th century. Although aluminum is a very common metal, comprising about 8% of the composition of the Earth's crust, it was only discovered in the 19th century due to its high affinity for oxygen and, therefore, its inability to be found in nature in its pure form. Pure aluminum was produced for the first time in laboratory conditions by Hans Christian Oerstedt in 1825, using the reduction process. Almost sixty years later, in 1886, two scientists—Hall and Héroult—each patented a process of smelting aluminum, independently of each other. The production of pure aluminum involves the dissolution of aluminum oxide (alumina—$Al_2O_3$) in molten cryolite and the electrolysis of a molten salt bath [1]. Additionally, Karl Jozef Bayer developed an efficient process for the production of alumina from bauxite in 1982, thereby completing the two-step so-called Bayer–Hall–Héroult process: refining bauxite to obtain alumina and smelting alumina to produce aluminum [2]. Bauxite is a naturally occurring, clay-like sediment that is the principal host for alumina. In general, bauxite is an ore formation consisting of a conglomerate of several minerals containing aluminum, with other valuable impurities. Iron oxides, such as $Fe_2O_3$ and FeO(OH), as well as clay minerals, kaolinite, rutile, titanium minerals, silica and anatase are among the contaminants. Bauxite is intended for aluminum production due to its high concentration of the aluminum-containing minerals Boehmite—AlO(OH), Gibbsite—$Al(OH)_3$ and Diaspore—AlO(OH).

From an industrial perspective, bauxite is a critical mineral from which alumina is extracted and subsequently used to produce primary aluminum.

Aluminum produced through the Baier–Hall–Heroult process is used to produce various primary casting alloys, which are applied in various industries. According to the available literature [3], the major global use of aluminum in 2021 was for the automotive and transport industry (29%), followed by construction (23%), foils and packaging (16%), electrical engineering and electronics (11%), machinery and equipment (7%), consumer goods (4%) and other applications (10%). The intensity of aluminum's use in the automotive industry and in transportation is increasing because of its unique properties (the combination of light weight and high strength), as well as its recyclability, which contributes to the reduction of greenhouse gas emissions. The global automotive industry has been going through drastic changes in the last several years [4]. One of the fundamental changes is its shift towards electrification. The shift is mainly triggered by new environmental regulations. Nevertheless, the idea to produce electric vehicles is not entirely new—as Figure 1 illustrates, it has existed for more than a century [5]. Several persons in Europe experimented, independently of each other, with electric cars during the second half of the 19th century (e.g., Porsche), while around the same time, in 1897, the Electric Carriage and Wagon Company in the USA introduced the first electric vehicles, which were commercially available [6,7]. However, gasoline-powered vehicles began to dominate the US market after 1920, mainly because of the development of a more comprehensive road infrastructure (including the implementation of petrol stations). Gasoline-powered vehicles could travel faster and further and were seen as superior to electric vehicles, which were slow and had limited driving range. The economic and oil crisis, air pollution and global warming are some of the factors that drastically changed the vehicle production strategy at the beginning of the 21st century, driven by more restricted environmental regulations. In the last few years, we have witnessed that most Original Equipment Manufacturers (OEMs) have introduced Electro Vehicle (EV) models or prototypes within their product portfolio, with increasing expectations for the coming years.

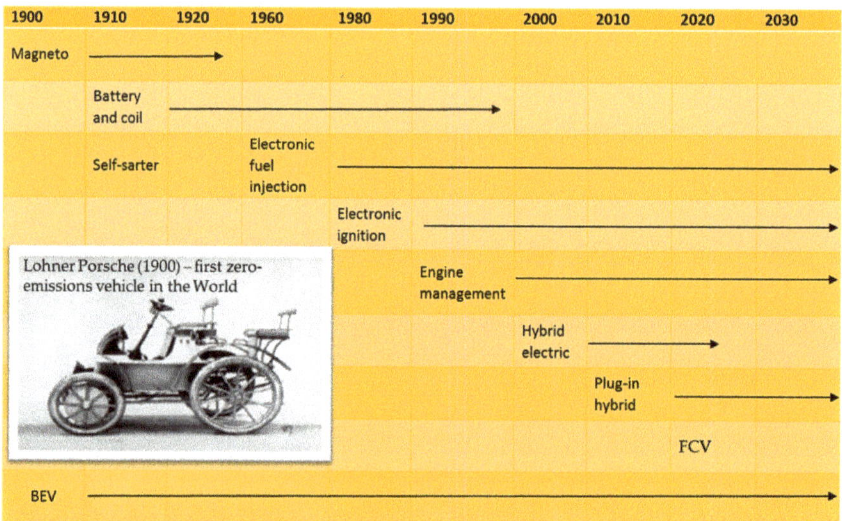


**Figure 1.** The historical perspective of road vehicle electrification [5].

The growing production of electric vehicles has caused the need to consider new properties of aluminum and its alloys in the production of e-mobility parts. Therefore, the goal of this study is to analyze what types of alloys, casting processes and heat treatment can be applied to the production of aluminum in casting plants in order to fulfill the required mechanical and thermophysical properties of electro automotive parts. Moreover, the intention is to highlight their new specific properties, such as dimensional stability, corrosion resistance, electromagnetic compatibility and crashworthiness, all of which are criteria that cast parts need to satisfy in order to be applied in the mobility industry. Electro automotive parts in this text are considered to fall under either e-mobility (e.g., battery, chasses, body, etc.) or e-housing (e.g., engine parts, stator, gearbox, etc.).

## 2. Aluminum Cast Alloys

Despite the great natural abundant of bauxite, as a primary raw material for aluminum cast alloy production, most of its end users are currently considering the use of secondary (recycled) aluminum for final production, due to very intensive energy consumptions and high $CO_2$ emission during primary aluminum production. The European Commission (EC) has bauxite listed as one of the critical raw materials (CRM) for 2020, counting on its strategic role in sustainable technologies and sectors from 2030 and 2050. The inclusion of bauxite in the EC's CRM list supports the role of Europe's aluminum industry in its transition to a green metallurgy. Thanks to its unique properties (low density, high thermal conductivity, non-toxicity, excellent corrosion resistance, as well as easy casting, machining and shaping), aluminum is recognized as the choice material for clean technology, especially in uses involving renewable energy, electrical systems, batteries, resource-efficient packaging, even energy-efficient buildings—and of course, aluminum alloys for cast automotive parts. Therefore, the sustainable production of aluminum from bauxite and/or bauxite residues is a very important and critical step for the future. Europe needs to develop a thriving and responsible obtaining strategy for bauxite as a CRM, considering recycling as required step on this road. The supply chain disturbances may be reduced by sustaining an irrepressible and autonomous European aluminum value chain, with the expectation of increased political and social attention. In this case, the aforementioned aluminum value chain can strengthen production capacities, regardless of whether it is primary or secondary aluminum.

The automotive industry has used aluminum–silicon cast alloys for years, primarily due to their outstanding mechanical properties and good casting characteristics in regard to producing body parts, transmission systems and motors. A recent study specially made for European Aluminum, which was conducted by Ducker Carlisle, revealed that the average amount of aluminum has increased by 18% in cars made in Europe, i.e., from 174 kg to 205 kg in the period of 2019–2022 [8,9]. The study predicts that this trend is set to continue, with the average aluminum content projected to increase to up to 256 kg per vehicle by 2030. According to the study, the trend towards light weighting and electrification in the automotive industry is driving a significant increase in aluminum content. This tremendous growth in aluminum usage in Battery Electric Vehicles (BEVs) is mainly attributed to its use in e-drive housing, battery pack housings, ballistic battery protection and cooling plates. Some of the cast aluminum alloys used in the production of Internal Combustion Engines (ICE) such as EN AC—47000 (AlSi12(Cu1)), EN AC—44300 (AlSi12(Fe)), EN AC—43500 (AlSi10MnMg) and EN AC—42100 (AlSi7Mg), will be also used in the production of parts for Battery Electric Vehicles (BEV). According to DIN 1706 [10], these alloys are standard alloys that can be acquired on the market. Definitively, the growth of the BEV market will have positive consequences for aluminum cast producers. Even though BEVs have fewer components and moving parts compared to traditional vehicles with internal combustion, aluminum will be used for some other parts (e.g., e-mobility, e-housings, etc.). Beside good mechanical and thermos-physical properties, these alloys need to satisfy some additional properties such as: dimensional stability, corrosion resistance, electromagnetic compatibility and crashworthiness. Three casting processes, presented in Table 1, have been identified as the most commonly used processes for the production of electro automotive parts: High Pressure Die Casting (HPDC), Low Pressure Die Casting (LPDC) and Core Package System (CPS). Table 1 also briefly summarizes the advantages of all three processes. More details about each casting process can be found in the literature [11,12].

**Table 1.** Review of some casting processes that can be used for the production of E-parts.

| Casting Process | | Advantages |
|---|---|---|
| High Pressure Die Casting (HPDC) | 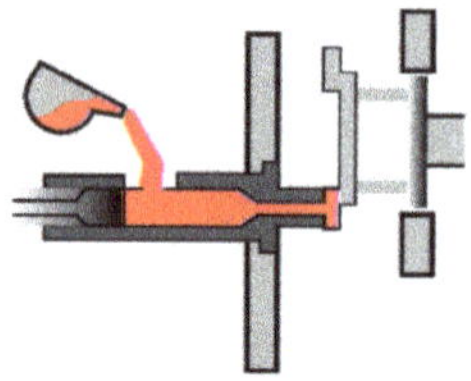 | • High production rates<br>• High level of automatization<br>• Good surface finish<br>• Economic in large quantity<br>• High flexibility in design<br>• Very good dimensional accuracy |
| Low Pressure Die Casting (LPDC) | 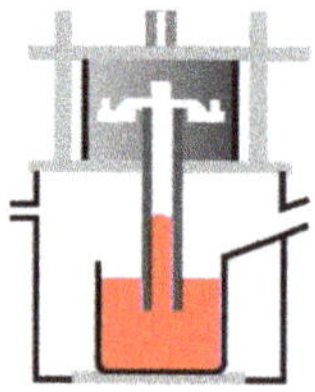 | • Fair production rate<br>• High quality of cast parts<br>• Moderate tooling cost<br>• High dimensional accuracy<br>• High material utilization—no need for feeders |
| Core Package System (CPS) |  | • Tranquil transfer of metal<br>• Minimal content of oxides and non-metallic inclusions<br>• High integrity and dimensional accuracy<br>• Reduced fettling time and no machining required.<br>• High strength and ductility of cast products |

## 3. Mechanical Properties

Electro automotive components are generally relatively uncomplex parts compared to cylinder heads or engine blocks. According to OEM, these alloys need to have a low residual stress level, a high dimensional stability, corrosion resistance, good electromagnetic compatibility and moderate mechanical properties in as-cast condition. These include yield strength (YS) $\geq$ 140 MPa, ultimate tensile strength (UTS) $\geq$ 180 MPa and elongation at fracture (A) between 1 and 2%. Table 2 shows typical mechanical properties of selected alloys taken from DIN 1706 [10] for die pressure casting process. According to Table 2, all selected alloys, even as-cast, achieved the expected mechanical properties (UTS and A).

**Table 2.** As-cast mechanical properties of die-cast selected alloys.

| Alloy | As-Cast Mechanical Properties | | |
|---|---|---|---|
| | UTS, MPa | YS, MPa | A, % |
| AlSi9Cu3(Fe) | 240 | 140 | <1 |
| AlSi10Mg | 240 | 140 | 1 |
| AlSi10MgMn | 250 | 120 | 5 |
| AlSi12(Fe) | 240 | 130 | 1 |
| AlSi12Cu1(Fe) | 240 | 140 | 1 |

Recently, some OEMs have begun to ask for solutions of BEV engines with very high torque (>1000 Nm). To reach this, e-housing and mobility parts need to provide significantly higher strength: YS > 240 MP and elongation: A = 2–3%. It means that the mechanical properties of future e-vehicle parts need to be improved through increasing strength and elongation at the same time. The dilemma of how to improve the strength of presently used alloys without sacrificing elongation is well known from the literature [13]. Figure 2 illustrates this well-known dilemma. As Figure 2 shows [13], applying various strengthening strategies such as: foreign particle reinforcement (blue closed squares), grain refinement (black closed circles), alloying (open squares) and optimized casting (green closed triangles), the strength of A356 alloy can be improved with minimum impact on elongation. From the literature and from foundry experience, it is known that the strength of HPDC aluminum–silicon cast alloys applied to the production of electro automotive parts can be improved by proper selection of casting processes (different cooling rates), applying heat treatment processes (solid solution strengthening mechanisms) and the selection of appropriate alloying elements (second phase strengthening and/or grain refinement).

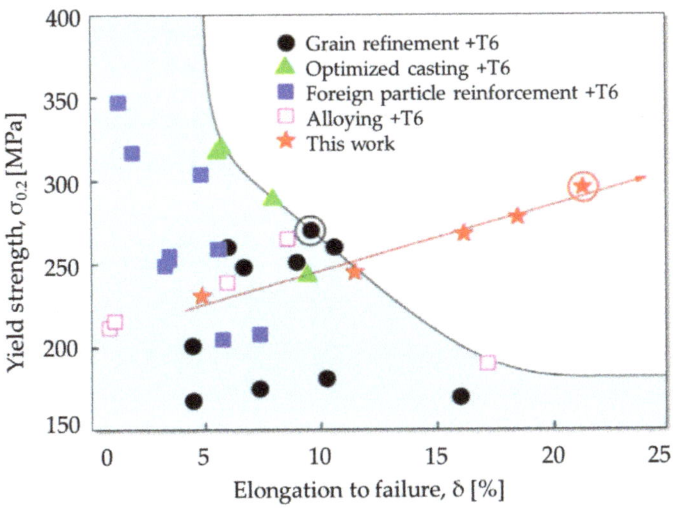


**Figure 2.** Strength–ductility trade off dilemma [13].

## 4. Improvement of Strength through Selection of Appropriate Heat Treatment Process

Cast aluminum alloys are lightweight structural materials that can be strengthened by further alloying. Additionally, some of the aluminum alloys which contain Cu, Mg and Zn can be further strengthened by heat treatment. The heat treatment of aluminum castings is carried out to change the properties of the as-cast alloys by subjecting each casting to a thermal cycle, or a series of thermal cycles. The conditions of thermal treatments of castings, as defined by EN 1706 [10], are as follows: F—as cast; T1—controlled cooling from casting and naturally aged; T4—solution heat treated and naturally aged; T5—controlled cooling from casting and artificially aged or over-aged; T6—solution heat treated, quenched (with water or air) and artificially aged; and T7—solution heat treated, quenched (with water or air) and over-aged (stabilized). The microstructure of materials is changed during the heat treatment process. Mechanical properties including strength, toughness, hardness, ductility and wear resistance are affected by the ensuing phase transformation [14]. As Figure 3 illustrates [15], by selecting the proper heat treatment process, strength or elongation can be significantly increased. By selecting T6 and T5 heat treatment states, the strength of the material can be increased, while by applying T7 and/or T4 its elongation can be improved. According to Figure 3, it should be considered that requested mechanical properties can be fine-tuned by selecting the optimal content of alloying elements, in this case magnesium. Besides improving mechanical properties, heat treatment also stabilizes the microstructure of cast parts and improves their thermal conductivity.

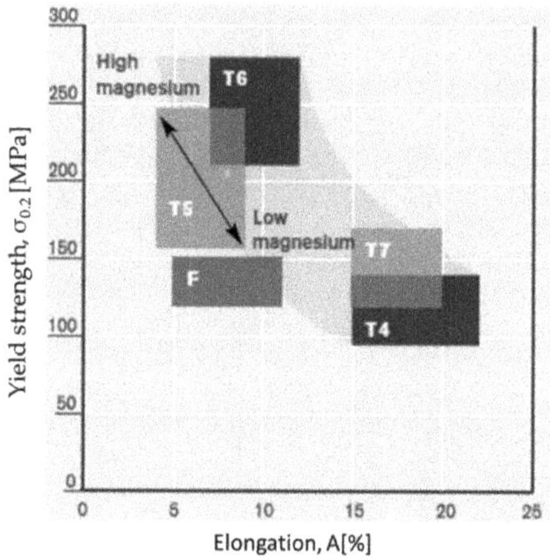


**Figure 3.** Impact of heat treatment processes on the elongation and strength of HPDC alloys [15].

It is well known from the literature that a combination of natural and artificial aging, as Figure 4 illustrates [16], can significantly improve strength at the expense of elongation. It appears that during natural ageing, small, coherent, finely distributed precipitates are formed, and during artificial ageing at higher temperatures, rather few, but coarse, incoherent particles precipitate. In addition, a combination of the two precipitation variants leads to the highest strengths in the casting parts [17].

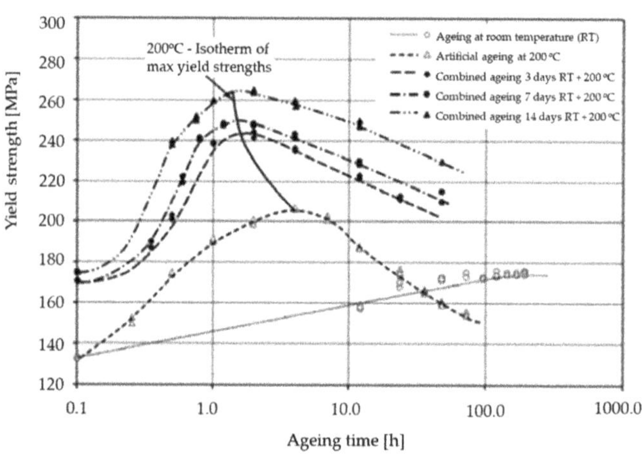


**Figure 4.** Increases in the yield strength through a combination of natural and artificial ageing at 200 °C; the isotherm shows the influence of natural ageing time on the maximum strength [16].

## 5. Improvement of Strength through Selection of Appropriate Alloying Elements

The effects of alloying elements on aluminum alloys include precipitation hardening (age hardening); dispersion strengthening; solid solution hardening; grain refinement; alteration of metallic and intermetallic phases; grain growth suppression at high temperatures; and so on. The mechanical properties of cast aluminum alloy components are determined largely by the shape, size and distribution of precipitated particles in the aluminum matrix. The impact of major and minor alloying elements on the mechanical properties (UTS, YS and A) of cast aluminum parts have been summarized in Table 3. More information regarding the impact of each alloying element on the mechanical properties of cast aluminum alloys can be found in the corresponding references listed in Table 3.

**Table 3.** A summary of the effects of some alloying elements on the mechanical properties of cast aluminum alloys.

| Alloying Element | | UTS | YS | A | References |
|---|---|---|---|---|---|
| Major | Si | ↗ | ↗ | ↘ | [18] |
| | Cu | ↗ | ↗ | ↘ | [19] |
| | Mg | ↗ | ↗ | ↘ | [20] |
| | Mn | Insignificant | Insignificant | ↗ | [21] |
| | Zn | ↗ | ↗ | ↘ | [22] |
| Minor | Fe | ↘ | ↘ | ↘ | [23] |
| | Pb | Insignificant | Insignificant | Insignificant | [24] |
| | Sn | ↘ | ↘ | ↗ | [25] |
| | In | ↗ | ↗ | ↗ | [26] |
| | Be | ↗ | ↗ | ↗ | [27] |
| | Cr | ↗ | ↗ | ↘ | [28] |
| | Ni | ↗ | ↗ | ↘ | [29] |
| | Zr | ↗ | ↗ | ↘ | [30] |
| | Mo | ↗ | ↗ | ↘ | [31] |
| | Sc | ↗ | ↗ | ↘ | [32] |
| | La | ↗ | ↗ | ↘ | [33] |

A literature review has shown that there are commercial HPDC alloys (e.g., AlSi7Mg, AlSi10MgMn) available on the market which can be utilized to create electro automotive parts with those essential characteristics. The mechanical properties of these alloys can be further improved by selecting a proper heat treatment process (T5, T6 and/or T7), which can adjust the strength and elongation of electro automotive parts produced using selected HPDC alloys. Selected HPDC alloys can further improve their strength characteristics by alloying with some major (Mg, Mn, Zn and Cu) and minor (Zr, Mo, Sc, La, etc.) alloying elements. The content of added elements needs to be optimized for each product to reach required mechanical, thermophysical, corrosion and other needed properties [34,35]. Additional consideration needs to be taken when using minor alloying elements. Their prices are significantly high and they are, therefore, sometimes not suitable for commercial application.

## 6. Dimensional Stability

In aluminum cast parts, the dimensional stability (irreversible casting growth) plays a significant role. In general, any precipitation of aluminum alloys during ageing is accomplished by the sample volume's growth, whereas dissolution during solid solution is followed by its size decreasing. This volume variation is caused by the precipitation of different phases (e.g., $Mg_2Si$, $Al_2Cu$, $Al_5Cu_2Mg_8Si$, and so on) during natural or artificial aging. Alloys with higher copper and magnesium contents are more prone to irreversible growth. This is mainly due to the Θ, Θ′ and Θ″ metastable copper-rich phases (the pre-stage of the $Al_2Cu$ stable intermetallic phase), which precipitate during ageing. Applying T5, T6 or T7 temper conditions causes those metastable phases to become more stable, and no significant further irreversible growth is expected during service thereafter.

Figure 5 shows that temperature and time strongly affect the dimensional stability of the proposed alloys. Lower test temperatures (lower working temperature conditions) for tested alloys Nr. 1 and Nr. 3 (Nr. 1—base alloy with higher content of copper and magnesium, and Nr. 3—alloy with lower content of copper and magnesium) resulted in lower irreversible casting growth rates. Measurements run at 100 °C for up to 400 h did not show any significant difference in the irreversible change of length for the two investigated alloys. As can be observed in Figure 5, irreversible growth increases with increasing ageing temperature (operating temperature). Beside temperature and time parameters, the amounts of precipitated phases play a significant role on the irreversible growth. It is obvious from Figure 5 that alloy Nr. 3, with a lower content of copper and magnesium, has a considerably smaller irreversible change of length compared to alloy Nr. 1 independently of the test temperatures (125 and 150 °C). Therefore, irreversible growth could be a significant issue only at localized positions of electro automotive parts that are exposed to temperatures higher than 120 °C.

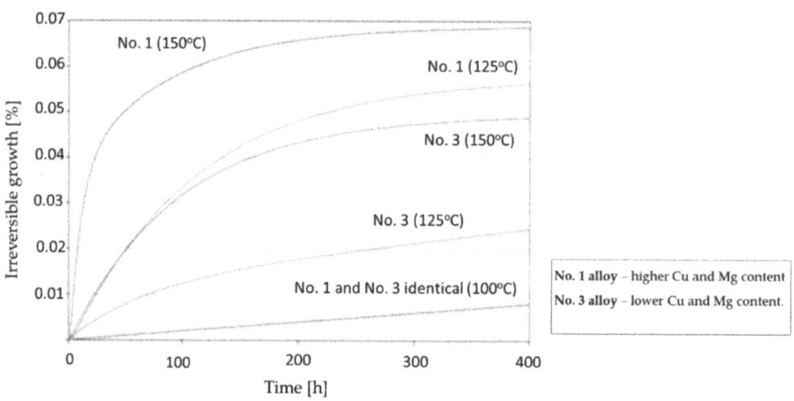


**Figure 5.** The impacts of temperature and time on irreversible growth during engine working conditions for an engine block [36].

## 7. Corrosion Resistance

Aluminum is a very active metal—it reacts immediately with oxygen if exposed to air, which naturally creates a passive adherent oxide layer on the surface, and its corrosion resistance is influenced by the passivity that this protective oxide layer creates [37]. In most environments, this thin layer instantly reforms after being damaged and keeps shielding the aluminum from corrosion. Corrosion happens when the film is taken off or damaged in a way that prevents self-repair. Aluminum is corrosion resistant in neutral solutions, but is highly susceptible to both basic and acidic solutions. In the pH range between 4 and 9, the amorphous protective oxide layer, with a thickness of 2 to 4 nanometers, is formed in water or air [38–40]. One of the fastest metals to oxidize is aluminum due to its extremely electronegative potential. The naturally passive layer, however, makes aluminum behave as a very stable metal, especially in oxidizing media like air and water. An aluminum alloy's ability to resist corrosion depends on both metallurgical and environmental factors [41–43]. Corrosion is influenced by metallurgical factors such as composition (see Figure 6), heat treatment (proper temper selection), mechanical working and the presence of impurities [39]. These influence the microstructure, which determines the type of attack and whether localized corrosion takes place.

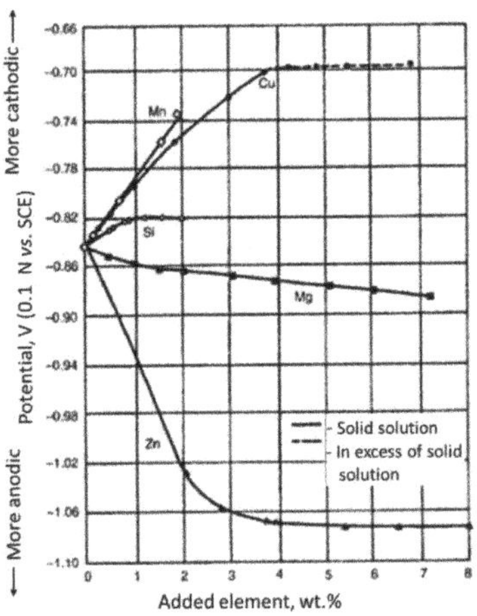


**Figure 6.** The effects of aluminum's major alloying elements on the electrolytic solution's potential [39].

Localized corrosion in aluminum alloys used in commercial engineering is typically brought on by microscopic flaws, like non-metallic inclusions or insoluble intermetallic particles with a size range of between several microns and a nanometer [40,44]. In addition, aluminum alloying elements also have some impacts on the corrosion. Elements such as silicon, manganese and copper can be precipitated in the form of intermetallic, in which form they are more cathodic (does not corrode, but stimulates corrosion in the adjacent matrix zone), while zinc and magnesium in intermetallic form are more anodic (stimulating corrosion). Therefore, aluminum alloys suitable for the production of electro automotive parts should have a lower copper content [40]. In the Al-Si-Cu alloys, precipitation of $Al_2Cu$ at the grain boundaries ($Al_2Cu$ is electrochemically nobler than the Al-matrix) can deplete the copper in the areas close to the boundaries, turning them anodic in comparison to the centers of the grains and causing intergranular corrosion to happen quickly. Therefore, when using cast alloys containing copper, some OEM producers of electro automotive parts expect a corrosive attack, especially on sealing surfaces and subsequently on the electrical connections [43]. In operation, electro automotive parts reach lower temperatures compared to combustion engines, which significantly aggravates a corrosive attack. Also, the use of stronger encapsulation can cause condensation phenomena, which likewise strengthens a corrosive attack. Therefore, the use of alloys with lower copper content is required [36,37].

Since there are numerous factors that affect corrosion, aluminum's suitability cannot be determined solely by a particular product or environment. It is crucial to have in-depth knowledge of impurities, operating circumstances, operating part design, as well as alloy microstructure. The most valuable experience comes from past service applications that were successful. However, improvements in fabrication (alloy selection, melt cleanliness), heat treatments and the development of suitable methods of protection may result in the successful application of Al-Si alloys with different copper contents in many corrosive environments [36–44].

## 8. Electromagnetic Compatibility

The ability of a unit of equipment or a system to operate effectively in its electromagnetic surroundings, without bringing intolerable electromagnetic disturbance into that environment, is known as electromagnetic compatibility. In other words, electromagnetic compatibility is the ability to tolerate a specific degree of interference. Due to the increasing use of electronic devices (automotive applications, personal computers, communication tools, and so on), electromagnetic compatibility is becoming more important in the automotive industry. In cases where selected alloys applied during the production of electro automotive parts are not in compliance with electromagnetic compatibility, the practical impacts can be recognized in both the minor (e.g., cell phone interference) and major consequences (e.g., improper deployment of airbags). Therefore, it is important to recognize the potential hazards of applying various aluminum alloys for use in electro automotive parts (e.g., battery housing), as their emitted magnetic fields can compromise the function and control of vehicles. Generally, aluminum and its alloys are well-protected against electromagnetic fields. Aluminum has very low magnetic permeability ($\mu R = 1.0002$), classifying it as part of the group of materials with paramagnetic behavior. A material's magnetic permeability—the level of magnetization it develops in response to an applied magnetic field—is a measure of its capacity to support the generation of a magnetic field within itself. Contrary to aluminum, iron has very high magnetic permeability ($\mu R = 300$ to 10,000), indicating ferromagnetic behavior in this metal. Copper in aluminum alloys precipitates in the form of copper-rich $Al_2Cu$ phase, which reduces electromagnetic fields by absorbing electromagnetic waves. The low electromagnetic permeability of aluminum alloys does not result from the fact that copper is a diamagnetic element (diamagnetic element means that this type of element is not attracted by magnets), but from the fact that an eddy current is induced at the surface, which protects against the electromagnetic fields. In addition, the presence of copper increases the eddy current, protecting against even stronger electromagnetic fields. Therefore, regarding electromagnetic compatibility, aluminum alloys with copper are suitable for application in the production of electro automotive parts.

## 9. Crashworthiness

The capacity of a structure to safeguard its users during impact is known as crashworthiness. This is frequently tested when examining the safety of vehicles and aircrafts. When a vehicle is said to be crashworthy, it means that its structural integrity can withstand reasonable deceleration loads and still maintain enough survival space for its users. In the case of a collision, all produced electro automotive castings which are part of the body structure, such as battery housings, must deform plastically in a short period of time (milliseconds) to absorb the crash energy in a controllable manner without cracking. At the same time, aluminum alloys used for electro automotive parts should be light weight, having sufficient flexibility and torsion stiffness for proper application and handling. All previously mentioned alloys (EN AC—47000 (AlSi12(Cu1)), EN AC—44300 (AlSi12(Fe)), EN AC—43500 (AlSi10MnMg), EN AC—42100 (AlSi7Mg)) have satisfactory ductility, which can be, when needed, further improved through alloying (e.g., modification by adding magnesium or strontium) and/or appropriate heat treatment processes. The solid solution treatment is mostly responsible for crashworthiness, as it involves modifying and separating silicon particles, minimizing crack initiation. The addition of strontium into aluminum alloys modifies large, plate-like silicon particles with sharp edges into small ones with rounded shapes, thereby improving the crashworthiness of the alloys.

## 10. Conclusions

Cast aluminum alloys are extremely attractive prospects for the production of electro automotive parts due to their respectable properties, such as: good mechanical and thermophysical properties, great dimensional stability, acceptable corrosion resistance, known electromagnetic compatibility and predictable crashworthiness. It means that this market will need more aluminum in the future. There are several cast aluminum alloys

(EN AC—47000 (AlSi12(Cu1)), EN AC—44300 (AlSi12(Fe)), EN AC—43500 (AlSi10MnMg), EN AC—42100 (AlSi7Mg)) available in the world market that can be used for the mass production of electro automotive parts by applying the following casting processes: HPDC, LPDC and CPS. The mechanical properties of these alloys can be further improved through suitable heat treatment processes (T5, T6 and T7) and additional alloying with major and minor alloying elements. The main concerns in regard to the addition of alloying elements to existing aluminum alloys are related to their limited content and their effect on the corrosion of the cast parts. Moreover, the cost of some alloying elements may potentially be the limiting factor for their use in the production of electro automotive parts.

**Author Contributions:** Conceptualization, methodology, validation, formal analysis, supervision M.D.; writing—original draft preparation, M.D., S.M., S.S. and M.M.; writing—review and editing, M.D. and M.M. All authors have read and agreed to the published version of the manuscript.

**Funding:** During this study M.M. was supported by the Ministry of Science, Technological Development and Innovation of the Republic of Serbia [Grant No. 451-03-47/2023-01/200026].

**Data Availability Statement:** No applicable.

**Conflicts of Interest:** The authors declare no conflict of interest.

## References

1. Vračar, R.; Živković, Ž. *Extractive Metallurgy of Aluminum*; Naučna Knjiga: Belgrade, Serbia, 1993; p. 314.
2. Gow, N.N.; Lozej, G.P. Bauxite. *Geosci. Can.* **1993**, *20*, 9–16.
3. Michi, R.; Plotkowski, A.; Shyam, A.; Dehoff, R.; Baby, S. Towards high-temperature applications of aluminum alloys enabled by additive manufacturing. *Int. Mater. Rev.* **2022**, *67*, 298–345. [CrossRef]
4. Czerwinski, F. Current Trends in Automotive Lightweighting Strategies and Materials. *Materials* **2021**, *14*, 6631. [CrossRef] [PubMed]
5. Grauers, A.; Sarasini, S.; Karlström, M. Why electromobility and what is it? In *Systems Perspectives on Electro Mobility*; Chalmers University of Technology: Gothenburg, Sweden, 2013; pp. 10–21.
6. Nieuwenhuis, P. The challenge of decarbonizing the car. In *Paving the Road to Sustainable Transport, Governance and Innovation in Low-Carbon Vehicles*; Nilsson, M., Hillman, K., Rickne, A., Magnusson, T., Eds.; Routledge Studies in Ecological Economics; Routledge: London, UK, 2012. Available online: https://www.book2look.com/embed/9781136316609 (accessed on 23 August 2023).
7. Available online: https://natural-resources.canada.ca/our-natural-resources/minerals-mining/minerals-metals-facts/aluminum-facts/20510 (accessed on 23 August 2023).
8. Djurdjevic, M.B.; Feikus, F.J.; Fernandez Gutierrez, R. Invited lecture: Properties of cast aluminum alloys suitable for production of E-mobility components. In Proceedings of the Metallurgical and Materials Engineering Congress of South-East Europe (MMSEE), Belgrade, Serbia, 5–7 June 2019; p. 9.
9. Steglich, J.; Luisa Marzoli, L.; Zerbin, I. Cast alloys for electric vehicle powertrains. In Proceedings of the European Aluminium Conference by GDA TRIMET Aluminium SE, Düsseldorf, Germany, 25–26 November 2019; pp. 1–18. [CrossRef]
10. *EN 1706*; Aluminium and Aluminium Alloys—Castings. European Standard: Brussels, Belgium, March 1998; pp. 1–18. Available online: https://aluminiumtoday.com/news/aluminium-usage-in-cars-surges-as-automotive-industry-shifts-towards-electrification (accessed on 23 August 2023).
11. Jolly, M.R. Castings. In *Comprehensive Structural Integrity*; Milne, I., Ritchie, R.O., Karihaloo, B., Eds.; Elsevier Pergamon: Oxford, UK, 2003.
12. Kaufman, J.G.; Rooy, E.L. *Aluminum Alloy Castings—Properties, Processes and Application*; Chapter—Heat Treatment of Aluminum Castings; AFS Schaumburg: Schaumburg, IL, USA, 2004; pp. 61–67, ISBN 0-87170-803-5.
13. Dang, B.; Zhang, X.; Chen, Y.Z.; Chen, C.X.; Wang, H.T.; Liu, F. Breaking through the strength-ductility trade-off dilemma in Al-Si based casting alloy. *Sci. Rep.* **2016**, *6*, 30874. [CrossRef] [PubMed]
14. Kim, S.-W.; Lee, S.-J.; Kim, D.-U.; Kim, M.-S. Experimental Investigation on Tensile Properties and Yield Strength Modeling of T5 Heat-Treated Counter Pressure Cast A356 Aluminum Alloys. *Metals* **2021**, *11*, 1192. [CrossRef]
15. *Primary Aluminum Alloys for Pressure Die Casting*, 4th ed.; Rheinfelden Alloys: Rheinfelden, Germany, 2016. Available online: https://rheinfelden-alloys.eu/wp-content/uploads (accessed on 23 August 2023).
16. Pabel, T.; Geier, G.F.; Rockenschaub, H.; Hopfinger, M. Improved mechanical properties of the high pressure die casting alloy AlSi9Cu3(Fe)(Zn) as a result of the combination of natural and artificial ageing. *Int. J. Mater. Res.* **2007**, *98*, 516–520. [CrossRef]
17. Kim, D.; Maeng, H.; Choi, Y.; Choi, H.; Lee, S.J. Constitutive Model of Triple-Step-Aged Al–Mg–Si Alloy Incorporating Precipitation Kinetics. *Met. Mater. Int.* **2021**, *27*, 4577–4585. [CrossRef]
18. Yuxian, L.; Hu, A.; Fu, Y.; Liu, S.; Shen, W.; Hu, H.; Nie, X. Al Alloys and Casting Processes for Induction Motor Applications in Battery-Powered Electric Vehicles: A Review. *Metals* **2022**, *12*, 216. [CrossRef]

19. Lumley, R.N.; O'Donnell, R.G.; Gunasegaram, D.R.; Kittel-Sherri, T.; Gershenzon, M.; Yob, A.C.; Polmear, I.J. The role of alloy composition in the heat treatment of aluminum high pressure die castings. *Mater. Sci. Technol.* **2008**, *26*, 1–10.
20. Caceres, C.H.; Davidson, C.J.; Wang, Q.G.; Griffiths, J.R.; Wang, Q.G. The Effect of Mg on the Microstructure and Mechanical Behavior of Al-Si-Mg Casting Alloys. *Metall. Mater. Trans. A Phys.* **1999**, *30*, 2611–2618. [CrossRef]
21. Abdulwahab, M. The Effect of Chromium and Manganese on the Mechanical Properties and Corrosion Resistance of Al-Si-Fe Alloy in 0.5M HCl Solution. Ph.D. Thesis, Department of Metallurgical Engineering Ahmadu Bello University, Zaria, Nigeria, August 2007.
22. Ammar, H.R.; Moreau, C.; Samuel, A.M.; Samuel, F. Influences of alloying elements, solution treatment time and quenching media on quality indices of 413-type Al–Si casting alloys. *Mater. Sci. Eng. A* **2008**, *489*, 426–438. [CrossRef]
23. Wang, L.; Makhlouf, M.; Apelian, D. Aluminum Die Casting Alloys—Alloy composition, microstructure, and properties-performance relationship. *Int. Mater. Rev.* **1995**, *40*, 221–238. [CrossRef]
24. Mohamed, A.M.A.; Samuel, F.; Samuel, A.M.; Doty, H.W. Effects of Individual and Combined Additions of Pb, Bi, and Sn on the Microstructure and Mechanical Properties of Al-10.8Si-2.25Cu-0.3Mg Alloy. *Metall. Mater. Trans. A Phys.* **2009**, *40*, 240–254. [CrossRef]
25. Mao, H.; Bai, X.; Song, F.; Song, Y.; Jia, Z.; Xu, H.; Wang, Y. Effect of Cd on Mechanical Properties of Al-Si-Cu-Mg Alloys under Different Multi-Stage Solution Heat Treatment. *Materials* **2022**, *15*, 5101. [CrossRef] [PubMed]
26. Akopyan, T.K.; Belov, N.A.; Letyagin, N.V.; Milovich, F.O.; Fortuna, A.S. Increased precipitation hardening response in Al-Si-Cu based aluminum casting alloy with In trace addition. *Mater. Today Commun.* **2021**, *27*, 102410. [CrossRef]
27. Yang, C.Y.; Lee, S.-L.; Lee, C.-K.; Lin, J.-C. Effects of Be and Fe on the mechanical and corrosion behaviors of A357 alloys. *Mater. Chem. Phys.* **2005**, *93*, 412–419. [CrossRef]
28. Yaro, S.A.; Aigbodion, V.S. The Effect of Chromium and Manganese Addition on the Corrosion of As-cast Al-Si-Fe-Cu Alloy System in Caustic Soda. *Eurasian Chem.-Technol.* **2008**, *10*, 79–84. [CrossRef]
29. Li, G.J.; Guo, M.-X.; Wang, Y.; Zheng, C.-H.; Zhang, J.-S.; Zhuang, L.-Z. Effect of Ni addition on microstructure and mechanical properties of Al–Mg–Si–Cu–Zn alloys with a high Mg/Si ratio. *Int. J. Miner. Metall. Mater.* **2019**, *26*, 740–751. [CrossRef]
30. Rahimian, M.; Amirkhanlou, S.; Blake, P.; Ji, S. Nanoscale Zr-containing precipitates; a solution for significant improvement of high-temperature strength in Al-Si-Cu-Mg alloys. *Mater. Sci. Eng. A* **2018**, *721*, 328–338. [CrossRef]
31. Ifeanyi, O.B.; Nkem, N.E.; Amaechi, A.F. Effect of Chromium and Molybdenum on the structure and mechanical properties of Al-Si alloys obtained by metal-mold casting. *J. Sci. Eng. Res.* **2016**, *3*, 383–389.
32. Emadi, D.; Prasada Rao, A.K.; Mahfoud, M. Influence of scandium on the microstructure and mechanical properties of A319 alloy. *Mater. Sci. Eng.* **2010**, *527*, 6123–6132. [CrossRef]
33. Ahmad, R.; Asmael, M.B.A. Influence of Lanthanum on Solidification, Microstructure, and Mechanical Properties of Eutectic Al-Si Piston Alloy. *J. Mater. Eng. Perform.* **2016**, *25*, 2799–2813. [CrossRef]
34. Choi, S.; Jeon, J.; Seo, N.; Son, S.B.; Lee, S.J. Effect of Heating Rate on Microstructure and Mechanical Properties in Al 7055. *Met. Mater. Int.* **2021**, *27*, 449–455. [CrossRef]
35. Patarić, A.; Mihailović, M.; Marković, B.; Sokić, M.; Radovanović, A.; Jordović, B. Microstructure as an essential aspect of 7075 aluminum alloy quality influenced by electromagnetic field during continual casting process. *Chem. Ind.* **2021**, *75*, 31–37. [CrossRef]
36. Rohr, A. Dauerfestigkeitserhöhung und Reduktion des irreversiblen Gusswachstums bei Aluminium-Sandgusslegierungen. Ph.D. Thesis, Technischen Universität Berlin, Berlin, Germany, 20 January 2012.
37. Jacobs, M.H. Corrosion and Corrosion Protection. In *TALAT Lecture 1252, Interdisciplinary Research Centre in Materials;* The University of Birmingham: Birmingham, UK; European Aluminum Association: London, UK, 1999; p. 17.
38. Campbell, F.C. *Elements of Metallurgy and Engineering Alloys;* ASM International Materials Park: Novelty, OH, USA, 2008; pp. 502–508, ISBN 978-0-87170-867-0.
39. Davis, J.R. *Alloying: Understanding the Basics;* ASM International: Almere, The Netherlands, 2001; pp. 351–416. Available online: www.asminternational.org (accessed on 23 August 2023).
40. Davoodi, A. Mechanistic Studies of Localized Corrosion of Al Alloys by High Resolution In Situ and Ex-Situ Probing Techniques. Ph.D. Thesis, KTH Chemical Science and Engineering, Stockholm, Sweden, 2007. ISBN 978-91-7178-817-7.
41. Jegdić, B.; Bobić, B.; Stevanović, M.; Mihailović, M.; Daničić, D.; Kovačina, J.; Radojković, B. Resistance to pit formation and pit growth for different tempers of AA2024 aluminium alloy in presence of benzotriazole. *Met. Mater. Int.* **2022**, *26*, 1643–1653. [CrossRef]
42. Wang, Z.; Fu, B.; Wang, Y.; Dong, T.; Li, J.; Li, G.; Zhao, X.; Liu, J.; Zhang, G. Effect of Cu Content on the Precipitation Behaviors, Mechanical and Corrosion Properties of As-Cast Ti-Cu Alloys. *Materials* **2022**, *15*, 1696. [CrossRef] [PubMed]
43. Wang, Z.; Dong, L.; Hu, B.; Chen, B. The effect of Cu addition on corrosion resistance of Al-Si-Mg-Cr alloy. *Metals* **2023**, *13*, 795. [CrossRef]
44. Revie, R.W. *Uhlig's Corrosion Handbook*, 2nd ed.; John Wiley & Sons, Inc.: Hoboken, NJ, USA, 2011; Chapter 40.

metals

MDPI

Review

# Recent Status of Production, Administration Policies, and Low-Carbon Technology Development of China's Steel Industry

Yufeng Qiao and Guang Wang *

State Key Laboratory of Advanced Metallurgy, University of Science and Technology Beijing, Beijing 100083, China; qyf2568@163.com
* Correspondence: wangguang@ustb.edu.cn

**Abstract:** In 2023, China's crude steel production amount reached 1.019 billion tons, and the energy consumption of China's steel industry amount reached 561 million tons of coal. China's steel industry, with its dominant reliance on coal for energy and the primary use of blast furnaces and converters in production processes, as well as its massive output, has become the main field for achieving China's "carbon peaking" and "carbon neutrality" goals. Firstly, this article summarizes the current production status of the steel industry and the situation of carbon emissions in the steel industry. Secondly, it discusses the dual-carbon policies based on the national and steel industry levels and outlines the future directions for China's steel industry. Subsequently, it analyzes the current state of research and application of mature and emerging low-carbon technology in China's steel industry and details the low-carbon plans of China's steel companies using the low-carbon technology roadmaps of two representative steel companies as examples. Finally, the article gives policy suggestions for the further carbon reduction of China's steel industry. The purpose of this paper is to show the efforts and contributions of China's steel industry to the early realization of its "carbon peaking" and "carbon neutrality" goals.

**Keywords:** steel industry; carbon peaking; carbon neutrality; low-carbon technology; low-carbon policy

## 1. Introduction

The increase in greenhouse gas emissions, such as carbon dioxide and methane, caused by human industrial, energy, and other activities is the main cause of global warming. Global warming leads to climate problems, such as rising sea levels and increasing extreme weather, and indirectly causes conflicts and economic issues. China currently possesses the world's most comprehensive and largest-scale steel industry system, equipped with the world's most advanced equipment and technology [1]. This industry can provide the most abundant iron and steel products. As an important basic industry, the steel industry contributes to the material foundation and material security of national economic development. However, the steel industry has been under significant pressure due to its environmental impact [2]. The steel industry is a resource-intensive and energy-intensive industry, while it is a high-energy-consumption and high-emission industry [3]. China's crude steel production is about 1.019 billion tons, ranking first globally. The steel industry's carbon emissions account for 15% of China's carbon dioxide emissions, and it is the second largest emitting industry [4]. Therefore, the steel industry faces enormous pressure to reduce carbon emissions.

China announced it will strive to reach peak carbon dioxide emissions by 2030 and achieve carbon neutrality by 2060 [5]. As a major contributor to China's carbon emissions, the steel industry has received increasing attention recently. The Chinese government, large steel enterprises, universities, research institutions, and others have all carried out relevant work. On 24 December 2021, the Chinese government issued the "Working Guidance for

Carbon Dioxide Peaking and Carbon Neutrality in Full and Faithful Implementation of the New Development Philosophy" [6]. It pointed out the formulation of action plans for carbon peaking in the energy, steel, and other industries to accelerate low-carbon technological development in the industrial sector. On 20 January 2022, the Chinese government issued "The Guiding Opinions on Promoting the High-Quality Development of the Steel Industry" [7]. It pointed out that innovative development, total quantity control, low-carbon development and overall co-ordination are the basic principles to achieve high-quality development in the steel industry. On 18 November 2021, Baowu Steel Group established a Low-Carbon Metallurgy Innovation Center and completed the construction of a Hydrogen-enriched Carbonic oxide Recycling Oxygenate Furnace (HyCROF), providing support for the progress of low-carbon metallurgical technology [8]. In December 2022, through the implementation of the "Hegang Low-Carbon Product Plan", Hegang Group promotes the innovation of low-carbon technologies with upstream and downstream industries by using carbon reduction technologies such as biomass energy substitution and carbon capture utilization and storage (CCUS) [9]. In 2021, Capital Engineering & Research Incorporated was approved to establish the Low-Carbon Technology Research Institute (Beijing) of China Metallurgical Corporation, which provides a solid foundation for future low-carbon technology research and development. Northeastern University, University of Science and Technology Beijing, Chongqing University, North China University of Science and Technology, and other universities have established low-carbon metallurgical research departments based on their advantageous disciplines of metallurgy and focus on the key technologies for low-carbon steelmaking to help China's steel industry to achieve low-carbon development.

Although many scholars have discussed carbon emissions and low-carbon technologies in China's steel industry [10,11], few have discussed low-carbon policies for the steel industry and steel enterprise's carbon reduction technologies. To fill this gap, firstly, this article overviews the development status of China's steel industry. Secondly, it discusses the dual-carbon policies based on the national and steel industry levels. Subsequently, it analyzes the current state of research on low-carbon technology in China's steel industry and details the low-carbon plans of China's steel companies. Finally, the article gives suggestions for further carbon reduction in China's steel industry. It is aiming to provide a reference for China's steel industry to achieve carbon peaking and carbon neutrality as soon as possible.

## 2. Current Production Status of China's Steel Industry

### 2.1. Variations of Crude Steel Production Amount

China's crude steel production amounts from less than one-thousandth of the world's production amount when the People's Republic of China was established, to 53.9% of the global crude steel production amount by 2023; this growth rate sets a new historical record for humanity [12]. Before 2000, China's crude steel production amount showed relatively stable growth. However, starting in 2001, with China's accession to the World Trade Organization and the subsequent surge in trade, China's crude steel production amount began to surge. It increased from 150 million tons in 2001 to 1.019 billion tons in 2023, and the trend is shown in Figure 1 [13,14]. According to data from the National Bureau of Statistics of China, the country's steel exports amount to 91.201 million tons, accounting for 8.94% of the production amount in 2023. This indicates that China's steel production primarily aims to meet domestic demand, and the economic condition heavily influences the crude steel production level.

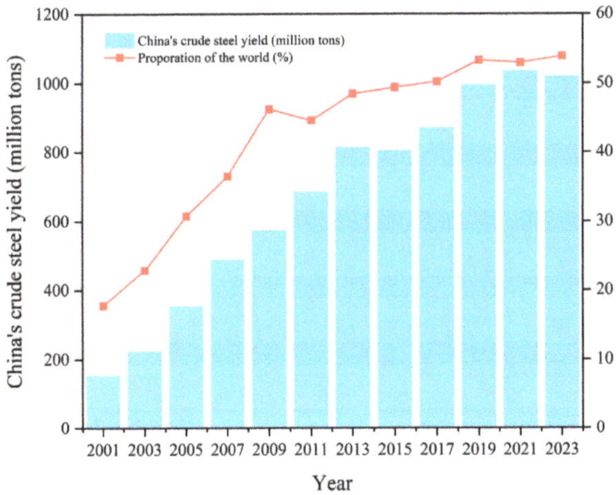


**Figure 1.** China's crude steel yield and the proportion of the world in recent years.

## 2.2. *Overview of Carbon Emissions and Challenges Ahead*

Currently, the carbon emissions from the Chinese steel industry account for approximately 15% of the country's carbon emissions. The primary use of blast furnaces and converters in production processes in China's steel industry contributes to the high level of carbon emissions because this process relies on coal and coke as the reducing agent and source of heat. The energy consumption of various major processes and carbon emission factors of the main raw materials are given in Table 1 and Figure 2 [15–17]. In recent years, the carbon dioxide emission from the Chinese steel industry has continued to increase, and the trend is shown in Figure 3. Until 2022, the carbon dioxide emissions from the Chinese steel industry had reached 1.823 billion tons [18–21].

**Table 1.** Process energy consumption of steel enterprises in 2023.

| | Sintering Process | Pellet Process | Coking Process | Blast Furnace Process | Converter Process | Rolling Process |
|---|---|---|---|---|---|---|
| kgce/t | 54.95 | 24.35 | 90.26 | 376.40 | −12.27 | 54.75 |

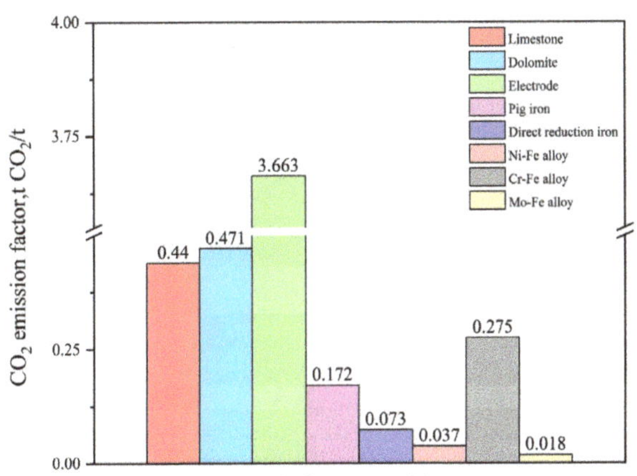


**Figure 2.** Carbon emission factors of main raw material.

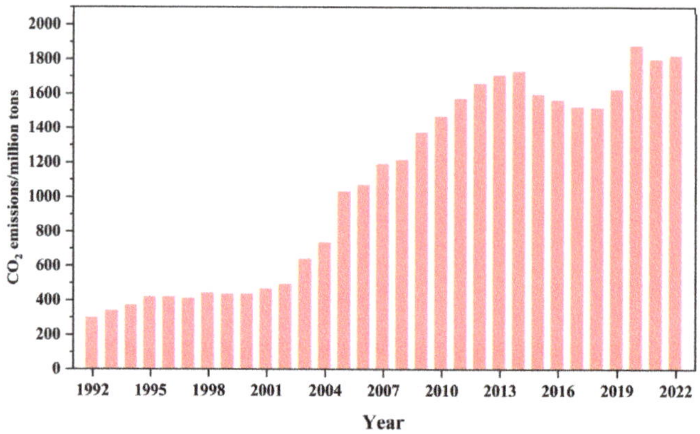


**Figure 3.** Changes in carbon emissions of China's steel industry in different years.

The Chinese steel industry faces several challenges in reducing carbon emissions [1,22–24]: (1) Increasingly stringent carbon emission regulations. Since 2021, the Chinese government and relevant departments have introduced a series of policies requiring steel enterprises to reduce carbon emissions. For example, "The Guiding Opinions on Promoting the High-Quality Development of the Steel Industry" emphasizes the need for the steel industry to achieve carbon peaking before 2030. (2) Imbalanced low-carbon technology levels among enterprises. Some leading steel companies have already reached or are close to world-class levels in terms of equipment and key technologies. However, some companies still have relatively outdated equipment and technologies, leading to high energy consumption and emissions. This imbalance hinders carbon reduction efforts in China's steel industry. (3) Lack of unified national low-carbon technology research and application plan. The Chinese steel industry has made significant progress in low-carbon technology recently. However, the development is still unbalanced. It is necessary to prioritize applying and developing universally applicable low-carbon technology. Mature and highly applicable low-carbon technology should be promoted and widely adopted as soon as possible.

### *2.3. Predictions on China's Steel Production Amount and Regional Distribution Changes*

Whether or not the carbon emission can peak quickly depends on the amount of crude steel production. Although the crude steel production amount in China started to decline in 2022 and may enter a sustained slow decline, however, to support the sustainable development of emerging economies and China's urbanization process, the steel industry will still maintain a large production amount in the future. It is projected that, by 2050, China's crude steel production amount will decrease to 600 million tons; the trend is shown in Figure 4 [25]. To realize the sustainable development of China's steel industry, we must focus on developing low-carbon and zero-carbon technology. It can contribute to national carbon neutrality goals by achieving carbon neutrality within itself.

The future production layout of crude steel in China will primarily depend on factors such as the place of crude steel consumption and ore production, the costs of hydrogen energy, transportation, and carbon capture and storage (CCS). Under deep decarbonization, the Chinese steel industry will mainly be concentrated in the northern regions by 2050 [26]. The crude steel production amount has significantly increased in Xinjiang and Qinghai due to the development of hydrogen and other renewable energy sources. On the other hand, regions that lack abundant renewable energy resources and have limited application of CCS technology will experience a noticeable reduction in the crude steel production amount. Therefore, for southern regions like Guangxi and Hunan, as well as major energy-consuming areas in eastern China, electric furnace steelmaking processes should be considered more.

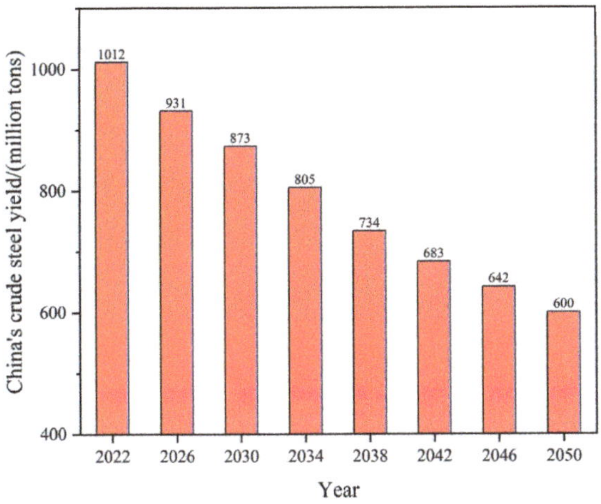


**Figure 4.** Forecast of China's crude steel production.

## 3. Low-Carbon Policies Related to China's Steel Industry

To actively implement the dual-carbon goals proposed by President Jinping Xi at the 75th United Nations General Assembly, China's government and ministries related to the steel industry have successively issued a series of relevant documents in recent years, which have provided guidance and principles for China's dual-carbon work.

### *3.1. National-Level Dual-Carbon Policies*

On 26 October 2020, the Chinese government issued "Proposals for Formulating the 14th Five-Year Plan (2021–2025) for National Economic and Social Development and the Long-Range Objectives through the Year 2035" [27]. It emphasized the need to accelerate the promotion of green development, strengthen legal and policy guarantees for green development, expedite the development and utilization of clean energy, reduce carbon emissions, support eligible regions in achieving carbon peaking at an earlier stage, and formulate an action plan to achieve the goal of carbon emissions peak before 2030.

On 24 October 2021, the Chinese government issued an "Action Plan for Carbon Dioxide Peaking before 2030" [28]. Over the 14th Five-Year Plan period, notable progress should be made in the adjustment and optimization of the industrial structure and the energy structure, and new improvements will be made in the R&D and broad application of low-carbon technology, making the goal to reach peak carbon emissions permeate the whole process and every aspect of economic and social development. Particular focus will be placed on implementing ten carbon-peaking actions: promote structural optimization for the steel industry and the substitution of clean energy, vigorously promote demonstrations of non-blast furnace technology, drive the application of advanced and appropriate technologies, and squeeze out all available potential for saving energy and cutting carbon emissions.

On 18 October 2021, the Chinese government released "Several Opinions on Strict Energy Efficiency Constraints to Promote Energy Conservation and Carbon Reduction in Key Areas" [29]. The document aims to gradually promote carbon reduction efforts in critical industries, fully utilizing the technological resources of domestic research institutes, universities, and leading enterprises. It seeks to expedite the development of cutting-edge low-carbon technology and promote the greening and high efficiency of key domestic enterprises' equipment. By 2025, the document requires the proportion of the standard-level production amount exceeding 30% in the steel industry through the implementation of energy conservation and carbon reduction actions.

### 3.2. China's Dual-Carbon Policies for Steel Industry

#### 3.2.1. Enhance Innovation Capabilities of Steel Industry

The Chinese government issued the "Guiding Opinions on Promoting the High-Quality Development of the Steel Industry". This includes the following: achieving breakthroughs in cutting-edge technologies, such as hydrogen metallurgy and non-blast furnace ironmaking; and highlighting the importance of enhancing the industry's innovation capabilities. Additionally, the document specifies the main research directions for key common technologies and all-purpose and specialized equipment, establishes co-operation mechanisms for key areas, and builds industrial alliances; and encourages qualified localities to plan and build regional innovation platforms in the steel sector and actively strive to create national-level innovation platforms.

#### 3.2.2. Strictly Control Iron and Steel Production Amount

To curb the continued increase in steel production amount, the "Guiding Opinions on Promoting the High-Quality Development of the Steel Industry" emphasizes the need to intensify efforts to prevent the blind implementation of smelting projects, and additional capacity in processing, casting, ferroalloys, etc., is prohibited; strictly avoid the resurgence of low-grade steel and strengthen penalties for violations; and increase the investigation of new steel production amounts in breach of regulations and use a legal way to effectively address the issue of overcapacity in the steel industry.

#### 3.2.3. Develop Electric Arc Furnace Steelmaking in an Orderly Manner

Compared to long-process steelmaking, electric arc furnace short-process steelmaking no longer requires ore as the raw material. It mainly consumes electricity as energy, significantly reducing $CO_2$ emissions. The "Guiding Opinions on Promoting the High-Quality Development of the Steel Industry" states that it encourages eligible enterprises to prioritize the transformation of electric arc furnace steelmaking. It also suggests establishing small- and medium-sized electric arc furnace steelmaking enterprises in central cities and urban clusters, and it can synergistically dispose of urban waste; and encourages regions with advantageous conditions to establish electric arc furnace steelmaking demonstration zones and explore and develop new technologies and equipment.

#### 3.2.4. Promote Restructuring of Steel Enterprises

The steel industry is an essential sector of China's national economy. The restructuring of steel enterprises can enhance the industry concentration, which, in turn, has several significant effects. Firstly, the increased concentration could reduce carbon emissions through regulated economies. Secondly, it helps in controlling the overall industry capacity. In 2020, the concentration ratio of China's steel industry was 39.2%, an increase of a five-percentage-point compared to the end of the 13th Five-Year Plan. However, the concentration level of Chinese steel enterprises remains relatively low, with a notable gap compared to other countries worldwide. Since 2005, the Chinese government has supported restructuring steel enterprises. A noteworthy example is the establishment of China Baowu Steel Group in 2016, which became the first steel conglomerate in China with an annual production amount exceeding 100 million tons.

#### 3.2.5. Deepen Co-operation in Low-Carbon Technology Development

The "Guiding Opinions on Promoting the High-Quality Development of the Steel Industry" deepens the promotion of low-carbon technology development as the main task; emphasizes supporting the construction of low-carbon metallurgical innovation alliances; establishes a data management system for carbon emissions in the production process, and formulate action plans for hydrogen metallurgy; and actively promotes the coupled development of steel, electric power, chemicals, and other industries.

## 4. Development Status of Low-Carbon Technology in China's Steel Industry

### 4.1. Quantitative Assessment of Carbon-Reducing and Energy-Saving Capacity of Different Low-Carbon Technologies

A quantitative assessment of the energy-saving and carbon-reducing capacity of steel industries is useful for the rational use of fuel, energy resources, and raw materials, and the improvement of the technological structure of production. A number of energy-modeling approaches have been used to analyze future trends and to assess the capacity for energy-saving and emissions reduction [30]. These can be categorized into two types: top-down and bottom-up models [31–33]. Most of the methods of assessment for low-carbon technologies in the steel industry use bottom-up analyses [34]. The advantage of the bottom-up analysis method is that it can simulate the development of technologies from a micro level, and forecast the energy-saving potential [35]. Some scholars divide 20 types of low-carbon technologies into four categories—coal-saving technology, electricity-saving technology, comprehensive energy-saving technology, and linkage technology—according to the energy-saving effect of the different technology on energy varieties, and construct a bottom-up model to analyze the carbon-reducing and energy-saving capacity under three different scenarios—the baseline scenario (BS), policy scenario (PS), and strengthened policy scenario (SPS). In terms of carbon reducing, the contribution of comprehensive energy-saving technology is highest; the emission reduction in 2030 is 129 million tons $CO_2$ ($MtCO_2$) and 130 $MtCO_2$, in the PS and SPS. In terms of energy saving, the contribution of coal-saving technology is highest; the ratios of coal saving in 2030 is 12.6% and 13.6%, in the PS and SPS [35].

### 4.2. Application Status of Typical Low-Carbon Technology

With the continuous development of China's steel industry in recent years, many carbon reduction technologies have reached leading levels and have been implemented in large domestic enterprises [36,37].

(1) Coal Moisture Control

In this technology, the coking coal is dried before being loaded into the coke oven through direct or indirect heating. The moisture content of the coking coal is stabilized at around 6% before the coking process begins. This technology significantly reduces the heat consumption during cooking, resulting a in lower energy consumption and notable emission reduction effects. Additionally, it helps achieve a more uniform particle size distribution of the coke and improves its quality. In China, the coal moisture control saves approximately 0.06 $GJ \cdot t^{-1}$ and reduces $CO_2$ emissions by about 1.47 $kg \cdot t^{-1}$. Coal moisture control technology has been widely adopted in major steel enterprises, such as Baosteel Group, Taiyuan Group, and Magang Group [38].

(2) Coke Dry Quenching

This method cools discharged coke from the oven with a cold, inert gas. It has significant advantages in recovering sensible heat from the discharged coke and optimizing operational conditions. It offers benefits such as improving coke quality and reducing energy consumption. In China, the coke dry quenching saves approximately 0.37 $GJ \cdot t^{-1}$ and reduces $CO_2$ emissions by about 42.54 $kg \cdot t^{-1}$. Over 330 sets of coke dry quenching equipment have been put into operation in China, with a processing capacity of approximately 46,000 $t \cdot h^{-1}$. The adoption rate of coke dry quenching in the national coking industry has exceeded 60% [39].

(3) Top Pressure Recovery Turbine

This technology uses gas expansion to drive a turbine, which generates electricity. The top pressure recovery turbine does not consume any fossil fuels, and it does not produce any pollutant gas emissions, resulting in significant emission reduction benefits. In China, the top pressure recovery turbine saves approximately 0.12 $GJ \cdot t^{-1}$ and reduces $CO_2$ emissions by about 22.66 $kg \cdot t^{-1}$. According to statistics from the China Iron and Steel

Industry Association, by 2010, over 600 blast furnaces were equipped with 597 sets of Top Pressure Recovery Turbine (TRT) equipment. For example, two 5500 $m^3$ blast furnaces with TRT technology in the Jingtang Steel Plant achieved a maximum power generation of 31,323 kW·h and a maximum daily average of 29,536 kW·h [40,41].

(4) Converter Dry Dedusting System

In this technology, the hot gas in the flue is cooled to about 200 °C, and then the cooled gas is treated with an electrostatic precipitator. This technology significantly reduces the emission of dust particles, achieves zero wastewater discharge, and offers advantages such as long service life and low operating costs [42]. In China, the converter dry dedusting system saves approximately 0.14 $GJ \cdot t^{-1}$ and reduces $CO_2$ emissions by about 5.77 $kg \cdot t^{-1}$. Major Chinese steel companies such as Baotou Steel, Taiyuan Iron and Steel, and Laigang Steel have used converter dry dedusting system technology since around 2005.

(5) Sinter Plant Heat Recovery

Sinter Plant Heat Recovery technology is primarily used to improve the energy utilization efficiency and reduce energy consumption during the sintering process. In the Chinese steel industry, the energy consumption of the sintering process is second, accounting for 10% to 20% of the production energy consumption, only to the ironmaking process. Therefore, the recovery and utilization of waste heat generated during the sintering process hold significant importance for the low-carbon development of the steel industry. In China, the sinter plant heat recovery technology saves approximately 0.35 $GJ \cdot t^{-1}$ and reduces $CO_2$ emissions by about 14.77 $kg \cdot t^{-1}$. There are 1200 sintering machines in operation in China, with a waste heat recovery utilization rate ranging from 20–30%; the sintering area is about 126,000 $m^2$ [43]

(6) Blast Furnace Injection of Natural Gas/Coke Oven Gas

In blast furnace injection of natural gas/coke oven gas, the reducing gases such as natural gas and coke oven gas were injected into the blast furnace, partially replacing the injection of coal and coke. Countries with abundant natural gas resources, such as Russia and the United States, have used natural gas injection in blast furnaces since the 1970s. The technology of blast furnace injection of natural gas is relatively mature. In 2019, at the No. 7 blast furnace (4800 $m^3$) in Indiana Harbor operated by ArcelorMittal in the United States, the natural gas injection rate was 50 kg/t, and the fuel rate was 480 kg/t [44]. Injecting hydrogen-rich gases into the blast furnace not only generates certain economic benefits through coke saving but also helps improve the company's economic performance by reducing $CO_2$ emissions. In China, the blast furnace injection of coke oven gas saves approximately 0.35 $GJ \cdot t^{-1}$, and reduces $CO_2$ emissions by about 14.77 $kg \cdot t^{-1}$. Jinnan Steel has conducted coke oven gas and hydrogen injection tests in three 1860 $m^3$ blast furnaces [45,46].

(7) High Percentage of Pellet/Lump Ore in BF Burden

The energy consumption of pellet ore in blast furnace iron production is half that of sinter ore. Pellet ore has advantages such as having a high grade, good reducibility, and lower emissions during the production process. Lump ore is a natural mineral and is cleaner [23]. Carbon emissions can be effectively reduced by increasing the proportion of pellet and lump ore in production. Ganbao Plant achieved a high level of 21.5% lump ore utilization in 2019 [47].

(8) High-efficiency Continuous Casting

High-efficiency continuous casting focuses on a high casting speed and is based on the production process being high-quality and defect-free. It aims to achieve high casting rates and high operation rates. The widespread adoption of high-efficiency continuous casting technology has increased the output of continuous casting machines and improved the surface and internal quality of cast slabs, leading to the production of defect-free billets. In China, the high-efficiency continuous casting saves approximately 0.39 GJ/t, and reduces

$CO_2$ emissions by about 27.49 kg/t. Chinese steel companies such as Wuhan Iron and Steel Group and Meigang Group have implemented various new technologies related to high-efficiency continuous casting in multiple stages of their production processes [48–50].

### 4.3. Research Status of New Low-Carbon Technology in China

In recent years, China has conducted extensive research on low-carbon technology [51–53], which has extensively promoted the development of low-carbon ironmaking technologies in the country. This article provides an overview of the development status of low-carbon technology in China, focusing on representative examples such as hydrogen-enriched carbonic oxide recycling oxygenate furnaces, coke oven gas zero-reforming shaft furnace direct reduction technology, HIsmelt technology, $CO_2$ steelmaking, and steel–chemical coproduction.

(1) Hydrogen-enriched Carbonic oxide Recycling Oxygenate Furnace

The Hydrogen-enriched Carbonic oxide Recycling Oxygenate Furnace is a modification of traditional metallurgical processes, where hydrogen replaces carbon, significantly reducing greenhouse gas emissions during the smelting process and, ultimately, achieving carbon neutrality for the entire smelting process. In July 2020, Baowu Group completed the construction of the world's first 430 $m^3$ hydrogen-enriched carbonic oxide recycling oxygenate furnace. In June 2021, the injection of the Ouye shaft furnace decarbonized gas at the tuyere was achieved. In July 2021, the hydrogen-rich metallurgical industrial experiments were completed, with a blast oxygen content of 50%, and the injection of coke oven gas and decarbonized gas at the tuyere was achieved. In 2022, the goal of full oxygen smelting was achieved, and the injection of decarbonized heating gas and coke oven gas was completed at the tuyere and furnace stack [54–56]. Baowu Group's experimental team has completed industrial production tests under conditions ranging from 35% oxygen to 50% oxygen [57]. In 2023, Baowu Group plans to construct a 2500 $m^3$ hydrogen-enriched carbonic oxide recycling oxygenate furnace.

(2) Coke Oven Gas Zero-Reforming Shaft Furnace Direct Reduction Technology

This technology primarily involves the reduction and deoxidation of oxidized pellets using a reducing gas consisting mainly of $H_2$ and CO, and accompanied by reactions such as the self-reforming of coke oven gas and carburization [58]. After self-reforming, the $H_2/CO$ ratio in the coke oven gas can reach above 8:1, becoming the highest hydrogen content in the gas-based shaft direct reduction process currently used. In December 2022, Hebei Iron and Steel Group completed the construction of the world's first project for the coke oven gas zero-reforming shaft furnace direct reduction technology. This project can reduce carbon emissions by approximately 70% compared to traditional metallurgical processes, reducing carbon emissions by approximately 0.5 kg/t [59].

(3) HIsmelt Technology

Compared to traditional blast furnace ironmaking, the HIsmelt process enables the direct use of non-coking coal powder and ordinary ore powder for smelting, eliminating the reliance on coking coal and simplifying the entire ironmaking process. The innovation of HIsmelt technology includes the use of a rotary kiln instead of a fluidized bed to reduce equipment failure rates, the replacement of refractory materials to optimize cooling systems and extend equipment lifespan, the development of high-life mixed injection lances, and the optimization of operations to achieve flexible iron output. Through continuous research and adjustments, in 2017, Shandong Molong achieved continuous production of HIsmelt and, in 2018, achieved 157 consecutive days of stable output [60].

(4) Carbon Dioxide Steelmaking Technology

Carbon dioxide has stability, non-burning-supporting properties, and the ability to undergo reactions at high temperatures. These characteristics enable $CO_2$ to play roles in stirring, cover protection, and dilution within the steelmaking process [61]. Beijing University of Science and Technology has invented several $CO_2$-related technologies, including

$CO_2$-$O_2$ dust reduction, $CO_2$ phosphorus removal, $CO_2$ nitrogen removal, $CO_2$ oxygen control, and $CO_2$ bottom blowing. In April 2022, the $O_2$-$CO_2$-CaO bottom-blowing technology was tested in an industrial setting at Shougang Jing-Tang Iron and Steel Company, achieving a $CO_2$ emission reduction of 30 kg/t [62].

(5) Steel–Chemical Coproduction

This technology, using blast furnace gas, converter gas, and coke oven gas of iron and steel enterprises as raw materials, involves extracting components such as carbon dioxide, carbon monoxide, and hydrogen through a series of purification and refinement processes, and then making methanol, ethanol, liquefied natural gas, and other high-value-added products after chemical processing [63,64]. Steel–chemical coproduction reduces carbon emissions from the source and achieves the coupling development of the chemical and steel industries. Numerous Chinese companies have begun the large-scale production of chemical products using exhaust gases, such as methanol, ethanol, ethylene glycol, and formic acid. In 2018, the first set of natural gas ethylene glycol projects was successfully tested by Xinjiang Tianying Petroleum Chemical [65].

### *4.4. Representative Steel Enterprises' Low-Carbon Technology Roadmap*

#### 4.4.1. Baowu Group's Low-Carbon Technology Roadmap

In 2021, Baowu Group released its low-carbon technology roadmap, as shown in Figure 5 [66]. It plans to achieve carbon peaking in 2023, can reduce carbon emissions by 30% in 2025, strives to achieve a 30% emission reduction in 2035, and strives for carbon neutrality in 2050. Its low-carbon goals are realized primarily by deploying six technologies: extreme efficiency, HyCROF, terminal manufacture, metallurgical resource recycling, hydrogen-based shaft furnace, and carbon recovery and utilization. From 2020 to 2035, the deployment of extreme efficiency, metallurgical resource recycling, terminal manufacture, and HyCROF will be completed. From 2035 to 2050, the deployment of hydrogen-based shaft furnace and carbon recovery and utilization technologies will be completed. The emission reduction targets for each technology category are as follows: extreme efficiency (3–5% reduction), metallurgical resource recycling (10–20% reduction), terminal manufacture (5–10% reduction), HyCROF (30–50% reduction), hydrogen-based shaft furnace (50–90% reduction), and carbon recovery and utilization (30–50% reduction).

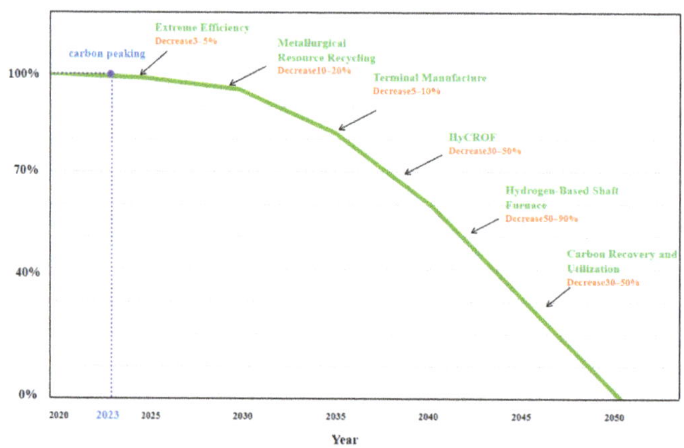


**Figure 5.** Baowu group's low-carbon technology roadmap.

#### 4.4.2. Hebei Iron and Steel Group's Low-Carbon Technology Roadmap

In 2022, Hebei Iron and Steel Group released its low-carbon technology roadmap, as shown in Figure 6 [67]. The roadmap outlines three phases of low-carbon development: "Peak Carbon Platform Period (2022–2025), Steady Decline Period (2025–2030), and

Deep Decarbonization Period (2031–2050)". Through implementing six major technology pathways, Hebei Iron and Steel Group aims to achieve a reduction in carbon emissions of over 10% compared to peak levels by 2025, over 30% by 2030, and carbon neutrality by 2050. The six major technology pathways include optimizing iron and steel resources, process optimizing and restructuring, improving system energy efficiency, optimizing energy structure, technological transformation for low-carbon, and collaborative decarbonization across industries. By 2050, it is projected that optimizing iron and steel resources will result in a 13% carbon reduction, process optimizing and restructuring will result in an 18% carbon reduction, improving system energy efficiency will result in a 34% carbon reduction, optimizing energy structure will result in an 11% carbon reduction, technological transformation for low-carbon will result in a 17% carbon reduction, and collaborative decarbonization across industries will result in a 7% carbon reduction [68].

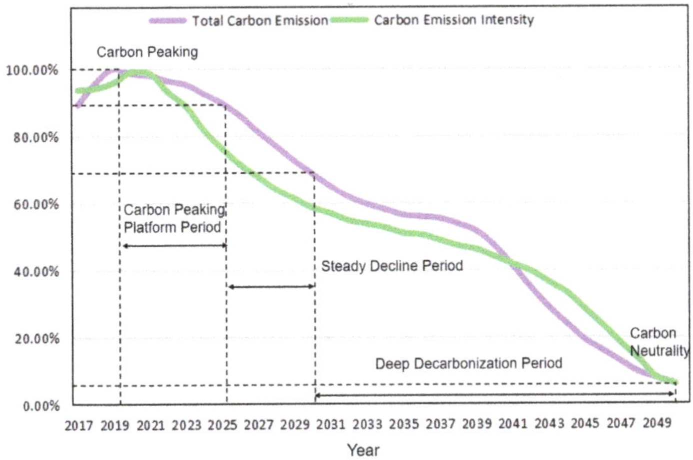


**Figure 6.** Hebei iron and steel group's low-carbon technology roadmap.

## 5. Suggestions for Further Reduction of Carbon Emission in China's Steel Industry

(1) Unify organization and implementation

On the one hand, we uniformly consider the demand for steel in the country's economic and social development, utilizing both domestic and international steel supply, under the premise of meeting the basic domestic steel demand, scientifically planning and forecasting of the trend of China's steel production, and scientifically leading China's steel materials and related industries towards high-end and volume reduction to ensure the economic competitiveness of the iron and steel industry. On the other hand, combined with the changes in China's scrap production and import volume, as well as the progress of decarbonization of the power system, we guide the development of electric arc furnace short-process steelmaking in a timely manner. The steel industry should formulate carbon neutrality targets in stages based on completing the above two works and clarifying the corresponding stages' key tasks and implementation paths.

(2) Establish national industrial experimental platform

Low-carbon technology research is a large, high-risk, and time-critical task that is difficult for individual organizations to afford. In the foreseeable future, steel materials will continue to serve as the essential foundation for developing the social economy. Therefore, the steel industry's low-carbon development is society's responsibility. It is necessary to provide financial and policy support for the research of low-carbon technology from the national level and to promote the study of low-carbon technology by establishing a national-level industrialization experiment platform.

(3) Select key technologies for research by process

Long processes currently dominate China's iron and steel smelting, but the proportion of short processes will significantly increase, eventually forming a complementary situation. Therefore, research on low-carbon technology in the steel industry should be carried out using long and short processes. The development of carbon reduction technologies based on long processes is of great significance to the carbon peaking and initial carbon reduction of China's steel industry, while the development of low-carbon technologies based on short processes is of decisive importance to realizing carbon neutrality in China's steel industry.

(4) Establish new research co-operation model

China's government attaches great importance to the goal of carbon peaking and carbon neutrality, so the relevant departments should play a leading role in the research of low-carbon technology. The research of low-carbon technology in the steel industry should be aimed at engineering applications, and enterprises, design institutes, and engineering companies in the leading position in the industry play an important role in the research of low-carbon technology. The research on low-carbon technology must rely on the proposal of original technology, which comes from a large amount of basic research. Therefore, it is inseparable from the active participation of universities and research institutes. The research work requires a large amount of workforce and material resources, and it is also essential to ensure the stability of funding channels and balance the relationship between inputs and benefits. Organizational models used in the European Union (EU) and Japan for metallurgical process development, such as Ultra Low $CO_2$ Steelmaking (ULCOS), $CO_2$ Ultimate Reduction in Steelmaking process by innovative technology for cool Earth 50 (COURSE50), etc., can be referenced, with efforts to mobilize more positive factors.

## 6. Conclusions

(1) China's steel industry faces significant pressure to reduce carbon emissions with its massive production, its dominant reliance on coal for energy, and the primary use of blast furnaces and converters in production processes. The Chinese steel industry has limited potential to reduce process energy consumption alone to achieve carbon peaking and carbon neutrality. It relies on technological innovation and continues investing in low-carbon technology research.
(2) China's steel industry prioritizes dual-carbon goals, and companies and research institutions are taking significant action to achieve them. However, the development of high-efficiency carbon reduction technologies in China is in its early stages and relatively behind compared to other countries. It has yet to reach the level of industrial application, and there is still a significant gap in obtaining world-class carbon reduction technologies.
(3) Many Chinese steel companies have not proposed clear dual-carbon plans, and more unified research activities are needed at the industry level. Companies should establish their dual-carbon plans, co-operate with relevant universities and research institutions in collaborative research, and strive to achieve dual-carbon goals as soon as possible.

**Author Contributions:** Conceptualization, Y.Q.; methodology, Y.Q.; formal analysis, Y.Q.; data curation, Y.Q.; writing—original draft, Y.Q.; writing—review and editing, G.W.; funding acquisition, G.W.; investigation, G.W.; supervision, G.W.; resources, G.W. All authors have read and agreed to the published version of the manuscript.

**Funding:** This research was funded by the National Natural Science Foundation of China (No. U1960205), the China Baowu Low-Carbon Metallurgical Innovation Foundation (No. BWLCF202101 and No. BWLCF202104), and the China Minmetals Science and Technology Special Plan Foundation (No. 2020ZXA01).

**Data Availability Statement:** No new data were created or analyzed in this study. Data sharing is not applicable to this article.

**Conflicts of Interest:** The authors declare no conflict of interest.

## References

1. Zhang, L.Q.; Chen, J. Discussion on Achieving "Carbon Peak" and Suggestions for Reducing Carbon in Iron and Steel Industry. *China Metall.* **2021**, *31*, 21–25+52.
2. Bataille, C.; Åhman, M.; Neuhoff, K.; Nilsson, L.J.; Fischedick, M.; Lechtenböhmer, S.; Solano-Rodriquez, B.; Denis-Ryan, A.; Stiebert, S.; Waisman, H.; et al. A Review of Technology and Policy Deep Decarbonization Pathway Options for Making Energy-Intensive Industry Production Consistent with the Paris Agreement. *J. Clean. Prod.* **2018**, *187*, 960–973. [CrossRef]
3. Yu, J.Y.; Xu, R.S.; Zhang, J.L.; Zheng, A.Y. A Review on Reduction Technology of Air Pollutant in Current China's Iron and Steel Industry. *J. Clean. Prod.* **2023**, *414*, 137659. [CrossRef]
4. Wang, X.Y.; Li, B.; Lv, C.; Guan, Z.J.; Cai, B.F.; Lei, Y.; Yan, G. China's Iron and Steel Industry Carbon Emissions Peak Pathways. *Res. Environ. Sci.* **2022**, *35*, 339–346.
5. Wang, Y.; Guo, C.H.; Chen, X.J.; Jia, L.Q.; Guo, X.N.; Chen, R.S.; Zhang, M.S.; Chen, Z.Y.; Wang, H.D. Carbon Peak and Carbon Neutrality in China: Goals, Implementation Path and Prospects. *China Geol.* **2021**, *4*, 720–746. [CrossRef]
6. Working Guidance for Carbon Dioxide Peaking and Carbon Neutrality in Full and Faithful Implementation of the New Development Philosophy. Available online: https://www.gov.cn/zhengce/2021-10/24/content_5644613.htm?eqid=dd0858d50015ab6900000004646e2a99 (accessed on 9 April 2024).
7. The Guiding Opinions on Promoting the High-Quality Development of the Steel Industry. Available online: https://www.gov.cn/zhengce/zhengceku/2022-02/08/content_5672513.htm (accessed on 9 April 2024).
8. Lin, A.C.; Yin, S.B.; Zhu, Y.; Zhao, H.Q. Summary on Iron and Steel Processes and Primary Investigation of Development Direction (Part two). *Yunnan Metall.* **2021**, *5*, 119–128+134.
9. Tong, S.; Ai, L.Q.; Hong, L.K.; Li, Y.Q.; Sun, C.J.; Wang, X.F. Development of hydrogen metallurgy in China and research progress of key technologies. *J. Iron Steel Res.* **2024**, *1*, 1–12.
10. Ren, L.; Zhou, S.; Peng, T.D.; Ou, X.M. A Review of $CO_2$ Emissions Reduction Technologies and Low-Carbon Development in the Iron and Steel Industry Focusing on China. *Renew. Sustain. Energy Rev.* **2021**, *143*, 110846. [CrossRef]
11. Zhang, J.S.; Shen, J.L.; Xu, L.S.; Zhang, Q. The $CO_2$ Emission Reduction Path towards Carbon Neutrality in the Chinese Steel Industry: A Review. *Environ. Impact Assess. Rev.* **2023**, *99*, 107017. [CrossRef]
12. Wang, H.F.; Ping, X.D.; Zhou, J.C.; Li, X.P.; Lu, L.J. Review and Prospect of Green Development for Chinese Steel Industry. *Iron Steel* **2023**, *58*, 8–18.
13. Dong, J.C.; Wang, X.Y.; Cai, B.F.; Wang, J.N.; Liu, H.; Yang, L.; Xia, C.Y.; Lei, Y. Mitigation Technologies and Marginal Abatement Cost for Iron and Steel Industry in China. *Environ. Eng.* **2021**, *10*, 23–31+40.
14. Huang, H.; Guan, M.K.; Wang, K.X.; Zhao, J.Y.; Yang, Q.Z. Research on the Coal Saving and Emission Reduction Potential of Advanced Technologies in China's Iron and Steel Industry. *Energy Sustain. Dev.* **2024**, *78*, 101373. [CrossRef]
15. Bei, L.; Gao, Y.W.; Wang, A.N.; Yang, W.; Wang, B.; Liu, H.T.; Sun, S.L. Analysis of Energy Consumption of Main Production Processes in Iron and Steel Enterprise. *Energy Res. Manag.* **2023**, *4*, 175–180.
16. Wei, Y.L.; Huang, S.L.; Yu, Z.W.; Wu, Z.Y.; Cheng, T.; Long, H.M.; Wann, W. Development and Application of Key Synergistic Technologies Continuously Reducing Energy Consumption in Sintering Process. *Sinter. Pelletizing* **2023**, *4*, 110–118+126.
17. *GB/T 32151.5-2015*; Greenhouse Gas Emissions Accounting and Reporting Requirements. Standards Press of China: Beijing, China, 2015.
18. Shangguan, F.Q.; Liu, Z.D.; Yin, R.Y. Study on Implementation Path of "Carbon Peak" and "carbon Neutrality" in Steel Industry in China. *China Metall.* **2021**, *31*, 15–20.
19. Zhu, R.; Wei, G.S.; Zhang, H.J. Research and Prospect of EAF Steelmaking with Near-Zero Carbon Emissions. *Iron Steel* **2022**, *57*, 1–9.
20. Wei, R.F.; Zheng, X.T.; Li, J.X.; Jiao, L.L.; Mao, X.M.; Xu, H.F. Reaction Behavior of Burden above Soft Melting Zone in Carbon Cycle Oxygen Blast Furnace. *Iron Steel* **2023**, *58*, 69–81.
21. Steady Implementation of Carbon Reduction in the Steel Industry. Available online: http://www.news.cn/energy/20230417/695fdc5079504549a7b96306e2034bd6/c.html (accessed on 20 September 2023).
22. Dong, H.B.; Liu, Y.; Wang, L.J.; Li, X.C.; Tian, Z.L.; Huang, Y.X.; McDonald, C. Roadmap of China Steel Industry in the Past 70 Years. *Ironmak. Steelmak.* **2019**, *46*, 922–927. [CrossRef]
23. Wang, G.; Zhang, H.Q.; Su, B.X.; Ma, J.C.; Zuo, H.B.; Wang, J.S.; Xue, Q.G. The Current Situation of Carbon Emission and Carbon Reduction in Chinese Steel Industry. *Ind. Miner. Process.* **2021**, *50*, 55–64.
24. Wang, Y.J.; Zuo, H.B.; Zhao, J. Recent Progress and Development of Ironmaking in China as of 2019: An Overview. *Ironmak. Steelmak.* **2020**, *47*, 640–649. [CrossRef]
25. Ren, M.; Lu, P.T.; Liu, X.R.; Hossain, M.S.; Fang, Y.R.; Hanaoka, T.; O'Gallachoir, B.; Glynn, J.; Dai, H.C. Decarbonizing China's Iron and Steel Industry from the Supply and Demand Sides for Carbon Neutrality. *Appl. Energy* **2021**, *298*, 117209. [CrossRef]
26. Jiang, K.J.; Xiang, P.P.; He, C.M.; Feng, S.B.; Liu, C.Y.; Tan, X.; Chen, S.; Dai, C.Y.; Deng, L.C. Impact Analysis of Zero Carbon Emission Power Generation on China's Industrial Sector Distribution. *Glob. Energy Internet* **2021**, *4*, 5–11.

27. Proposals for Formulating the 14th Five-Year Plan (2021–2025) for National Economic and Social Development and the Long-Range Objectives through the Year 2035. Available online: http://politics.people.com.cn/n1/2021/0313/c1001-32050444.html (accessed on 7 April 2024).
28. Action Plan for Carbon Dioxide Peaking before 2030. Available online: https://www.gov.cn/zhengce/content/2021-10/26/content_5644984.htm (accessed on 7 April 2024).
29. Several Opinions on Strict Energy Efficiency Constraints to Promote Energy Conservation and Carbon Reduction in Key Areas. Available online: https://www.gov.cn/zhengce/zhengceku/2021-10/22/content_5644224.htm (accessed on 7 April 2024).
30. Wen, Z.G.; Meng, F.X.; Chen, M. Estimates of the Potential for Energy Conservation and $CO_2$ Emissions Mitigation Based on Asian-Pacific Integrated Model (AIM): The Case of the Iron and Steel Industry in China. *J. Clean. Prod.* **2014**, *65*, 120–130. [CrossRef]
31. Böhringer, C.; Rutherford, T.F. Integrated Assessment of Energy Policies: Decomposing Top-down and Bottom-Up. *J. Econ. Dyn. Control.* **2009**, *33*, 1648–1661. [CrossRef]
32. Lin, Q.G.; Huang, G.H.; Bass, B.; Qin, X.S. IFTEM: An Interval-Fuzzy Two-Stage Stochastic Optimization Model for Regional Energy Systems Planning under Uncertainty. *Energy Policy* **2009**, *37*, 868–878. [CrossRef]
33. Suganthi, L.; Samuel, A.A. Energy Models for Demand Forecasting—A Review. *Renew. Sustain. Energy Rev.* **2012**, *16*, 1223–1240. [CrossRef]
34. Ko, F.K.; Huang, C.B.; Tseng, P.Y.; Lin, C.H.; Zheng, B.Y.; Chiu, H.M. Long-Term $CO_2$ Emissions Reduction Target and Scenarios of Power Sector in Taiwan. *Energy Policy* **2010**, *38*, 288–300. [CrossRef]
35. Tan, X.H.; Li, H.; Guo, J.X.; Gu, B.H.; Zeng, Y. Energy-Saving and Emission-Reduction Technology Selection and $CO_2$ Emission Reduction Potential of China's Iron and Steel Industry under Energy Substitution Policy. *J. Clean. Prod.* **2019**, *222*, 823–834. [CrossRef]
36. Zhang, Q.; Zhang, W.; Wang, Y.J.; Xu, J.; Cao, X.C. Potential of Energy Saving and Emission Reduction and Energy Efficiency Improvement of China's Steel Industry. *Iron Steel* **2019**, *54*, 7–14.
37. Zhu, T.Y.; Liu, X.L.; Wang, X.D.; He, H. Technical Development and Prospect for Collaborative Reduction of Pollution and Carbon Emissions from Iron and Steel Industry in China. *Engineering* **2023**, *31*, 37–49. [CrossRef]
38. Guo, Y.L.; Hu, J.H.; Zhou, W.T. Development and Application Status of CMC in China. *Mod. Chem. Ind.* **2019**, *36*, 8–12.
39. Pan, D.; Xu, Q.X. Research of Peak Carbon Emissions and Carbon Reduction Pathways in Coking Industry. *Fuel Chem. Process.* **2023**, *54*, 5–7.
40. Han, Y.J.; Cao, Y.J.; Tao, Y.Z. Process Optimization and Productive Practice of Shougang Jingtang No.1 Blast Furnace. *Metall. Power* **2010**, *4*, 20–21+25.
41. Zhou, J.C.; Zhang, C.X.; Han, W.G.; Li, X.P.; Shangguan, F.Q. Application Status and Development Trend of TRT Technology in Chinese Steel Industry. *Iron Steel* **2015**, *50*, 26–31. [CrossRef]
42. Zhang, F.M.; Zhang, D.G.; Zhang, L.Y.; Han, Y.J.; Cheng, S.S.; Yan, Z.H. Research and Application on Large BOF Gas Dry Dedusting Technology. *Iron Steel* **2013**, *48*, 1–9+43.
43. Dong, H.; Wang, A.H.; Feng, J.S.; Zhang, Q.; Cai, J.J. Progress and prospect in sintering waste heat resource recovery and utilization technology. *Iron Steel* **2014**, *49*, 1–9.
44. Abdel Halim, K.S. Effective Utilization of Using Natural Gas Injection in the Production of Pig Iron. *Mater. Lett.* **2007**, *61*, 3281–3286. [CrossRef]
45. Gao, J.J.; Zhu, L.; Ke, J.C.; Huo, X.F.; Qi, Y.H. Industrialized application of hydrogen-rich gas injection into blast furnace of Jinnan Steel. *Iron Steel* **2022**, *57*, 42–48.
46. Xiao, X.W.; Wang, G.; Li, M.M.; Zhou, Z.P.; Zhao, Y.J. Analysis of carbon reduction strategy based on hydrogen-rich or carbon-containing gas used for smelting reductant. *China Metall.* **2023**, *33*, 121–127.
47. Song, S.T.; Zhao, Z.J. High Lump Iron Ore Proportion Production Practice of Hanbao Iron. *Gansu Metall.* **2021**, *43*, 11–13.
48. Zhang, J.J.; Mao, X.P.; Wang, C.F.; Zhu, W.J. Development and expection on high-efficiency producing technology of steelmaking in TSCR line. *Iron Steel* **2019**, *54*, 1–8.
49. Zhu, M.Y. A Study of Transport Phenomena and Key Technologies for High-Speed Continuous Casting of Steel. *Iron Steel* **2021**, *56*, 1–12.
50. Zhu, M.Y. Some Considerations for New Generation of High-Efficiency Continuous Casting Technology Development. *Iron Steel* **2019**, *54*, 21–36.
51. Gao, J.J.; Qi, Y.H.; Yan, D.L.; Wang, F.; Xu, H.C. Development Path and Key Technical Problems of Low Carbon Ironmaking in China. *China Metall.* **2021**, *31*, 64–72.
52. Wang, X.D.; Hao, L.Y. Analysis of Modern Ironmaking Technology and Low-Carbon Development Direction. *China Metall.* **2021**, *31*, 1–5+18.
53. Zhang, Q.; Shen, J.L.; Ji, Y.M. Analysis of Carbon Emissions in Typical Iron- and Steelmaking Process and Implementation Pathresearch of Carbon Neutrality. *Iron Steel* **2023**, *58*, 173–187.
54. Liu, H.X.; Wu, L.M.; Zhao, L.; Cui, Y.; Wang, S.Q.; Si, H.L.; Zhou, Y.N.; Zhou, H.; Xue, J.C. Analysis on $CO_2$ emission characteristics and treatment technologies in iron and steel industry. *Sinter. Pelletizing* **2022**, *47*, 38–47.
55. Sui, P.; Ren, B.L.; Wang, J.S.; Wang, G.; Zuo, H.B.; Xue, Q.G. Current Situation and Development Prospects of Metallurgical By-Product Gas Utilization in China's Steel Industry. *Int. J. Hydrog. Energy* **2023**, *48*, 28945–28969. [CrossRef]

56. Xu, W.Q.; Fu, L.; Yang, Y.; Tian, Q.L.; Li, C.Q.; Wang, Y.; Zhu, Y.; Hao, R. Research progress in low carbon smelting technology of blast furnace gas cycle coupling carbon capture. *Energy Environ. Prot.* **2023**, *37*, 175–184.
57. Development Status and Suggestions of Hydrogen Metallurgy in China's Steel Industry. Available online: http://www.csteelnews.com/xwzx/jrrd/202109/t20210910_54739.html (accessed on 20 September 2023).
58. Wang, X.D.; Zhao, Z.L.; Li, C.M.; Yang, Y.G. Hydrogen Metallurgy Engineering Technologies Based on Zero Reforming of Hydrogen Enriched Coke Oven Gas. *Iron Steel* **2023**, *58*, 11–19.
59. HBIS Group the World's First! The First Phase of HBIS's 1.2 Million Ton Hydrogen Metallurgy Demonstration Project Was Fully Completed. Available online: https://www.hbisco.com/en/news/group/t101/2075 (accessed on 20 September 2023).
60. Zhang, G.Q.; Zhang, X.F.; Han, J.Y.; Liu, D.Y. The research process and production application of HIsmelt reduction technology. *Hebei Metall.* **2022**, *5*, 39–43.
61. Zhu, R.; Wang, X.L.; Liu, R.Z. Recent Progress and Prospective of Application of Carbon Dioxide in Ferrous Metallurgy Process. *China Metall.* **2017**, *27*, 1–4+10.
62. Zhu, R.; Ren, X.; Xue, B.T. Research Progress of Ultimate Carbon Emission in BOF Steelmaking Process. *Iron Steel* **2022**, *58*, 1–10.
63. Guo, Y.H. Current station and tendency of purification and upgrading of blast furnace gas. *J. Iron Steel Res.* **2020**, *32*, 525–531.
64. Guo, Y.H.; Zhou, J.C. Current situation and future outlook of steel chemical co-production in China. *China Metall.* **2020**, *30*, 5–10.
65. Yang, Q.; Yang, Q.C.; Xu, S.M.; Zhu, S.; Zhang, D.W. Technoeconomic and Environmental Analysis of Ethylene Glycol Production from Coal and Natural Gas Compared with Oil-Based Production. *J. Clean. Prod.* **2020**, *273*, 123120. [CrossRef]
66. Baowu Group Low-Carbon Technology Roadmap and Zero Carbon Industry. Available online: https://www.baowugroup.com/glcmia/detail/246785 (accessed on 20 September 2023).
67. Wang, X.D.; ShangGuan, F.Q.; Xing, Y.; Hou, C.J.; Tian, J.L. Research on the low-carbon development technology route of iron and steel enterprises under the "double carbon" target. *Chin. J. Eng.* **2023**, *45*, 853–862. [CrossRef]
68. Hebei Iron and Steel Group's Low-Carbon Technology Roadmap. Available online: http://heb.chinadaily.com.cn/a/202203/02/ws621f63e9a3107be497a08c89.html (accessed on 20 September 2023).

*metals*

*Article*

# Importance of Changes in the Copper Production Process through Mining and Metallurgical Activities on the Surface Water Quality in the Bor Area, Serbia

**Radmila Marković [1,*], Vesna M. Marjanović [1,*], Zoran Stevanović [1], Vojka Gardić [1], Jelena Petrović [1], Renata Kovačević [1], Zoran Štirbanović [2] and Bernd Friedrich [3]**

1 Mining and Metallurgy Institute Bor, Alberta Ajnštajna 1, 19210 Bor, Serbia; zoran.stevanovic@irmbor.co.rs (Z.S.); vojka.gardic@irmbor.co.rs (V.G.); jelena.petrovic@irmbor.co.rs (J.P.); renata.kovacevic@irmbor.co.rs (R.K.)
2 Technical Faculty in Bor, University of Belgrade, Vojske Jugoslavije 12, 19210 Bor, Serbia; zstirbanovic@tfbor.bg.ac.rs
3 IME Process Metallurgy and Metal Recycling, Intzestraße 3, 52056 Aachen, Germany; bfriedrich@ime-aachen.de
* Correspondence: radmila.markovic@irmbor.co.rs (R.M.); vesna.marjanovic@irmbor.co.rs (V.M.M.); Tel.: +381-62-450762 (R.M.); +381-69-631736 (V.M.M.)

**Abstract:** This paper considers the impact of copper mining-influenced water and metallurgical wastewater on the surface water in the Bor area, Serbia. Sampling, realized through the four campaigns (2020–2021), confirmed that both types of water, discharged without appropriate treatment in the Bor River, had a signific impact on the concentration of metal ions, pH and electrical conductivity on the watercourse in the Bor area. The highest concentrations of the following metal ions, Cu—271 mg/L, As—25,991 μg/L, Ni—13,856 μg/L, Cd—2627 μg/L, and Pb—2855 μg/L, were registered in the metallurgical wastewater samples. After changes occurred in the copper production process by stopping the discharge of untreated wastewater into the Bor River, the concentrations of monitored elements were drastically decreased. In the period 2022–2024, the concentration values for Cu, As and Pb ions were below the maximum allowable value, and the concentration values of Ni and Cd ions were also decreased. The values for pH and electrical conductivity were in the maximum allowable range. The return of wastewater to the copper production process would lead to both a reduction in the primary water consumption and reduction in the negative impact on the environment.

**Keywords:** metal ions; copper production; mining and metallurgy; environment; wastewater; surface water

## 1. Introduction

Mining and metallurgy are among the industries that have the biggest impact on the environment [1]. Dust from the open pit, overburden, and flotation tailing dumps, as well as gases from smelters and wastewater from electrolysis plants, are the main sources of air, water, and soil pollution. On the other side, the surface and underground water streams are endangered by mining-influenced water and industrial wastewater from metallurgical facilities.

One of the biggest still-working copper mines—including mines with surface exploitation, underground mines, facilities for mineral processing, and metallurgical plants for copper production—is located near the city of Bor, Serbia (Western Balkan). The Bor copper mine is located in the northeastern part of Serbia, about 11 km west of Bulgaria, and about 70 km south of Romania. The area of the Bor ore field occupies the central part of the Timok eruptive massif. This area is bounded on the west by Crni Vrh (1027 m), on the north by Mali Krš (920 m) and Veliki Krš (1148 m), and on the south side, the terrain is much lower and there is no distinct height. The first mining operations started in 1903 and have been continuously operating for over 120 years. During this period, very large amounts of different kinds of waste were generated. Copper mineralization is predominantly porphyry deposits, with a sulfur mineral associated with pyrites. Contact of pyrite with water and oxygen leads to generating the acid mine drainage (AMD) or acid rock drainage (ARD) forms [2].

It is estimated that about 750 million tons of mining waste, out of about 378 million tons of flotation tailings [3], is deposited near the city of Bor during the exploitation period. This waste contains hazardous and potentially toxic elements such as copper, nickel, arsenic, zinc, antimony, mercury, chromium, bismuth, and other pollutants [4]. Besides polluting the environment, these pollutants are very toxic to human health and can cause a large number of diseases and even death in cases of very high doses when acute poisoning occurs [5–12]. In the Waste Management Strategy of the Republic of Serbia for the period 2022–2031, mining waste from the exploitation of the mineral raw materials is classified into waste group 01 in the waste catalog [13]. During 2020, the ore and stone extraction sector produced 45,709,000 tons of waste, most of which was deposited in landfills [14]. The mining and metallurgy activities in the Bor area are the largest producer of waste.

Problems with mining waste are primarily related to the disposal of a large amount of flotation tailings and overburden. However, since environmental protection was not the focus in the past, these waste materials were not managed properly. The environment in the surroundings of the open pit mines was polluted to the extent that the land, as well as surface and underground water, was not for human consumption or agriculture. Also, the metallurgical facilities, located in the center of the city of Bor, caused a large amount of pollution, especially the gases from the smelter plant that polluted the air, but also other facilities such as wastewater from the electrolysis plant, which is an important source of water pollution. Monitoring of the environment has shown the contamination of air, soil and water with metals, inorganic and organic compounds, PM 10 and PM 2.5, etc. [15,16]. The great negative impact of mining and metallurgy activities on the surface water quality, especially close to the active and abandoned mines, needs some measures for reduction, remediation and elimination of the pollutants [17,18].

The Bor River, Krivelj River, Ravna River, Bela River, and Veliki Timok River are the main rivers of the Bor area watercourse. There are also small creeks and mine-influenced water, along with industrial wastewater from metallurgical processes for copper production flowing into some of the mentioned rivers. As a consequence of the mining and metallurgy activities, some of these rivers have become very polluted mostly by metal ions, such as Pb, Zn, Cd, Ni, Se, As, Fe, etc. The Bor River is one of the most polluted rivers in Europe [19,20]. To find measures and solutions for the reduction, rehabilitation, and elimination of the polluting elements, it is necessary to understand the consequences of more than a century of continuous mining and metallurgical activities in the Bor mining area.

The results for characteristic parameters of the surface water, mine-influenced water, and metallurgical wastewater in the Bor area are presented in this paper. The main objective was to determine the difference in the surface water quality before and after the cessation inflow of the metallurgical wastewater and AMD water from different sources into the Bor River. In case of an inflow of metallurgical wastewater and AMD water into the Bor River, the data were collected during the four sampling campaigns, realized in the period from September 2020 to June 2021. Collecting data of the surface water quality in the Bor area was continued from October 2021 after stopping the metallurgical wastewater and AMD water from inflowing into the Bor River; data are available on the website of the city of Bor. All results were used for a comparative analysis of the surface water quality in the watercourses of the Bor area.

## 2. Materials and Methods

### 2.1. Study Area

The study area includes the surface water and water under the influence of the mining and metallurgy activities, and the starting point is the Bor River after merging the untreated municipal wastewater and AMD water from different locations. The other sampling points are located downstream towards the Danube River within a 40 km radius of the city of Bor. This area was chosen because mining and metallurgical activities have been going on for more than 120 years and they are the main sources of pollution [21].

The main sources of water pollution in the Bor area as well as the industrial zone, Krivelj; overburden (near the village of Oštrelj); flotation tailing dumps; metallurgical zone Bor; and open pits are presented in Figure 1.

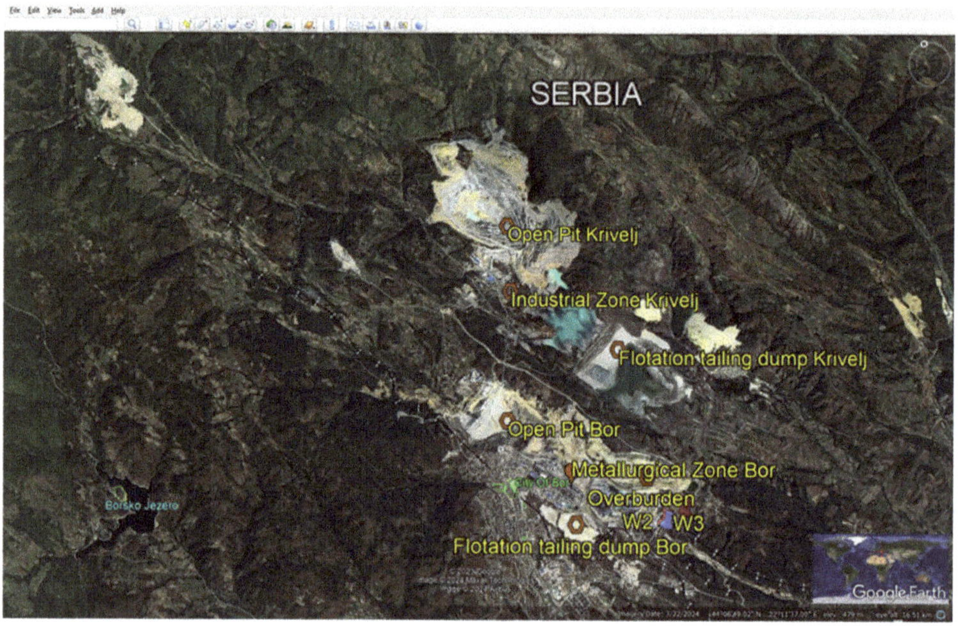


**Figure 1.** Sources of mining and metallurgical pollution of surface water.

The nine locations chosen to take samples for the physical–chemical characterization aim to define the concentration of Cu, As, Ni, Cd, Pb ions, pH and electrical conductivity (EC) of those samples (Figure 2). The Bor River is formed from the untreated municipal wastewater, AMD water from overburden deposited near the village of Oštrelj (W1), AMD water from flotation tailing dumps which are not in operation (W2) and metallurgical wastewater (W3), also discarded without previous treatment. Sampling of the Bor River (W4) was carried out on the point near the confluence with the Krivelj River (W5). Sampling of the surface water from the Krivelj River was carried out at the place where a significant influence of all wastewater from the study area was (mine, drainage, flotation). The Bela River (W6) was sampled after merging the Bor, Krivelj and Ravna rivers. The Ravna River (W7), which flows relatively far from the area of copper mining and metallurgy activities, was included for an analysis on the basis of literature data about pollution in the earlier exploitation period, realized by the Roman Empire [22].

The Timok River was sampled at two locations: a location marked as W8, out of influence of the mining and metallurgy activities from the study area (near the city of Zaječar), and a location marked as W9, the Veliki Timok River (near the village of Rajac), after inflow of the Bela River into the Timok River. The Veliki Timok River is a last tributary of the Danube River in the Republic of Serbia.

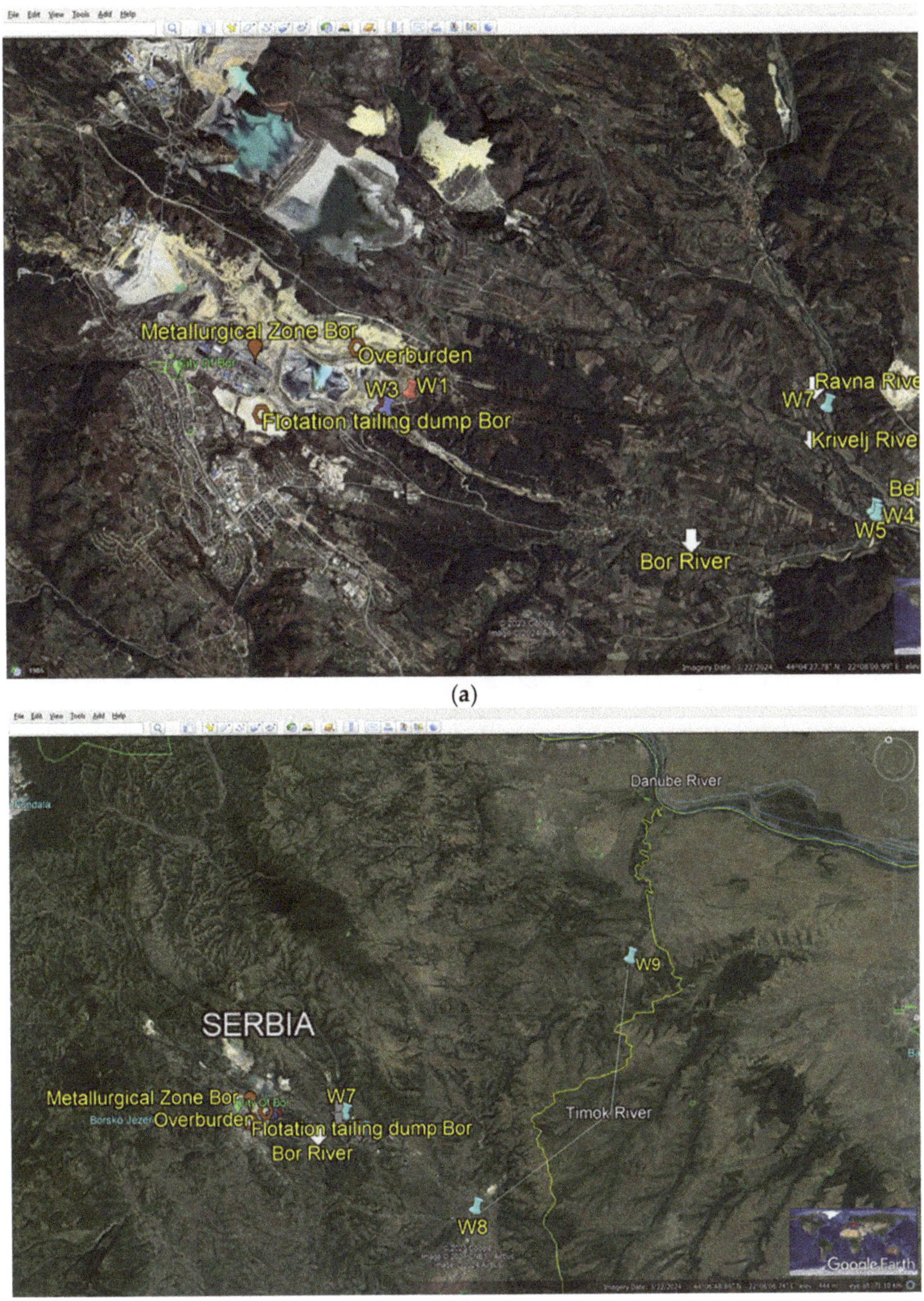


(a)

(b)

**Figure 2.** Water sampling locations: (**a**) locations under direct impact of mining and metallurgy activities; (**b**) all sampling locations.

Coordinates of water sampling locations are measured by a GPS device (Global Positioning System) Garmin, eTrex Vista® HCx (Olathe, Kansas City, MO, USA), and presented in Table 1.

**Table 1.** Coordinates of sampling locations.

| Water Sample Mark | Sampling Location | GPS Coordinates | |
|---|---|---|---|
| | | Lat. N | Long. E |
| W1 | AMD water from overburden | 44°03′46.65″ | 22°08′11.92″ |
| W2 | AMD water from flotation tailing | 44°03′47.73″ | 22°07′51.92″ |
| W3 | Metallurgical wastewater | 44°03′42.45″ | 22°07′53.22″ |
| W4 | Bor River | 44°01′46.81″ | 22°12′29.46″ |
| W5 | Krivelj River | 44°01′49.33″ | 22°12′28.92″ |
| W6 | Bela River | 44°01′37.65″ | 22°13′18.56″ |
| W7 | Ravna River | 44°02′39.69″ | 22°12′22.24″ |
| W8 | Timok River (out of impact the mining and metallurgy activities) | 43°56′40.30″ | 22°18′54.03″ |
| W9 | Veliki Timok River (after inflow of the Bela River into the Timok River) | 44°05′36.03″ | 22°34′02.60″ |

*2.2. Sampling Procedure*

A crucial factor in water sampling is to collect samples properly with the appropriate equipment so that the analytical results or field measurements will reflect the environmental conditions at the time of sampling. There is a possibility that errors are inadvertently produced during each sampling. This can lead to wrong results and bad conclusions.

Water samples (AMD water, industrial wastewater, and surface water) were sampled by the hand tools according to the standard sampling methods:

- Plastic containers from a high-density polyethylene (volume of 1 L), used for general water sample collection, are rinsed several times in the river or water that will be sampled before filling with a sample.
- An unfiltered water sample is poured into a plastic bottle of 50 mL without prior washing because 2.5 mL of $HNO_3$ conc. is present there. Exactly 50 mL of the sample is poured into a syringe, and after that, the bottle is sealed and labeled. Before use of a syringe, it should be rinsed three times with a water sample, and contents should be discharged.
- Samples are properly stored in the shipping containers to ensure samples are between 4 and 10 °C, and transported to the laboratories of the Mining and Metallurgy Institute Bor.
- All samples are properly stored from the time they arrive at the laboratory to disposal. Samples are refrigerated at 4 °C before analysis unless the method protocol indicates other storage conditions. In general, the shorter the time that elapses between sample collection and its analysis, the analytical results will be more reliable.
- For certain constituents and physical values, an immediate analysis in the field is required. On a sampling site, part of a water sample from a plastic container was poured into a vessel to measure the pH and electrical conductivity. The vessels for pH and electrical conductivity measuring were also three times rinsed with a water sample before performing the measure.
- All field measurement instruments are calibrated before starting sampling and once again after completing all of the sampling.
- At each sample location, the details relevant to the subsequent analysis and interpretation are entered in the lists for water sample collection and testing. The details in the list include all information necessary to assist in data interpretation and repeatability of sampling. Completion of the list starts in the field, during the sampling activities. The rest of the forms are completed step by step as samples are analyzed; the results will be obtained.

### 2.3. Physical Chemical Characterization and Statistical Analysis

Relevant international and European standards are used for prescribed appropriate methods for determining the metal ion concentration (Cu, As, Ni, Cd, and Pb), pH, and electrical conductivity.

The Coupled Plasma Optical Emission Spectroscopy analytical technique ICP-AES (model: SPECTRO ARCOS ICP-OES SPECTRO Analytical Instruments GmbH Bosch str. 10, 47533 Kleve, Germany) was used to determine the Cu ions. The Mass Spectrometry with Inductively Coupled Plasma analytical technique ICP-MS (model: Agilent 7700, Agilent Technologies Japan, Ltd. 9-1, Takakura-machi, Tokyo 192-8510, Japan) by the external calibration method was used to determine the As, Ni, Cd, and Pb ions.

The Certified Reference Material (CRM) was used for the quality control of chemical analysis in the laboratory conditions. An analysis of replicate test portions will indicate the repeatability and precisions of measurement.

Electrical conductivity and pH are measured on the field in duration of 10 min. Electrical conductivity is measured with EC1387, CHEMLAND (Stargard, Poland) and pH with WTW 7310, Ino Lab. (Xylem Analytics Germany Sales, GmBH&Co, KG, Weilheim, Germany).

Calculations of the average values and standard deviation of metal ion concentration of samples from the period September 2020–June 2021 were made by the Microsoft Excel 2016 Data Analysis Statistical functions and software package OriginPro 8 [23].

## 3. Results and Discussion

### 3.1. Characteristics of Wastewater and Surface Water in the Bor Area (Sampling Period September 2020–June 2021)

The results of more extensive research of the metal ion concentration, pH and electrical conductivity in wastewater, generated from copper mining and metallurgy activities and surface water from rivers, under the influence of mentioned activities, are presented in this paper. Therefore, the observed parameters in eastern Serbia or other areas with copper mining activities can represent a reference point for further research. The water samples, originating from different rivers, mining-influenced water, and metallurgical wastewater streams, are extremely enriched with toxic elements [24]. Sampling was performed during the four campaigns, in the period from September 2020 to June 2021, at defined locations (Table 1). It was aimed at determining the following parameters: pH, electrical conductivity, Cu, As, Ni, Cd, and Pb ion concentration.

According to the Serbian regulation on watercourse categorization, the observed rivers are categorized as the III and IV categories (Table 2). Also, Table 2 presents the maximum allowable values (MAVs) for characteristic parameters.

**Table 2.** MAVs according to the Serbian legislation for different river categories, adapted from Refs. [25,26].

| Parameters | River Category and MAVs | |
|---|---|---|
| | III Category (after Inflowing of the Bela River into the Timok River) until the Confluence of the Veliki Timok and Danube Rivers | IV Category (from the City of Bor to the Confluence of the Bela River and Timok River (Out of Impact of the Mining and Metallurgy Activities) |
| pH | 6.5–8.5 | 6.5–8.5 |
| Electrical conductivity, μS/cm | 1500 | 3000 |
| Cu, mg/L | 0.5 | 1 |
| As, μg/L | 50 | 100 |
| Ni, μg/L | 34 | 34 |
| Cd, μg/L | 0.6 | 0.9 |
| Pb, μg/L | 14 | 14 |

3.1.1. Concentration of the Cu, As, Ni, Cd and Pb Ions

Figures 3–7 present the obtained values for the concentration of the Cu, As, Ni, Cd and Pb ions during each sampling campaign, in the period from September 2020 to June 2021. Also, all graphs present the MAVs for the corresponding river category.

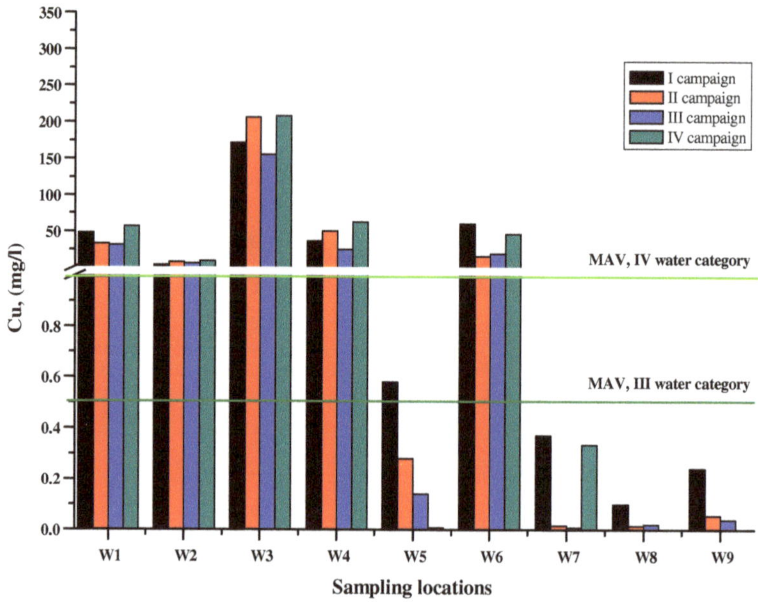


**Figure 3.** Concentration of Cu ions in the realized sampling campaigns at defined locations.

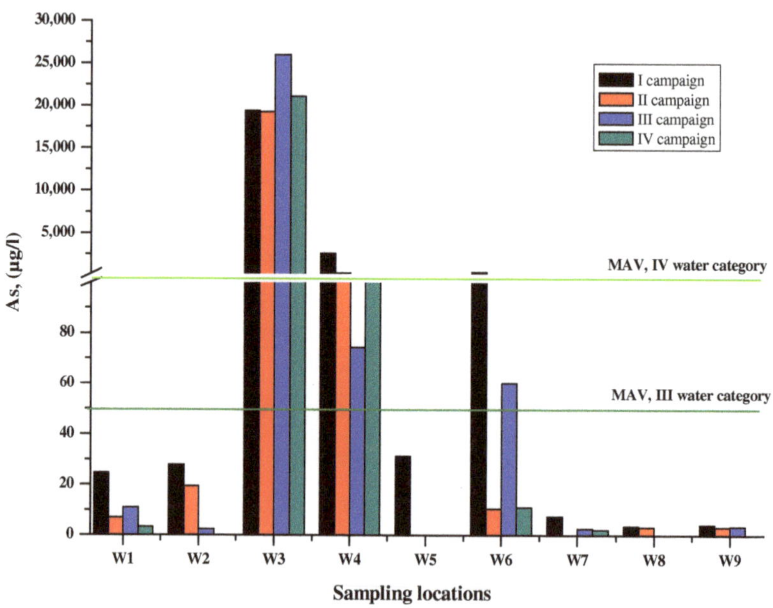


**Figure 4.** Concentration of As ions in the realized sampling campaigns at defined locations.

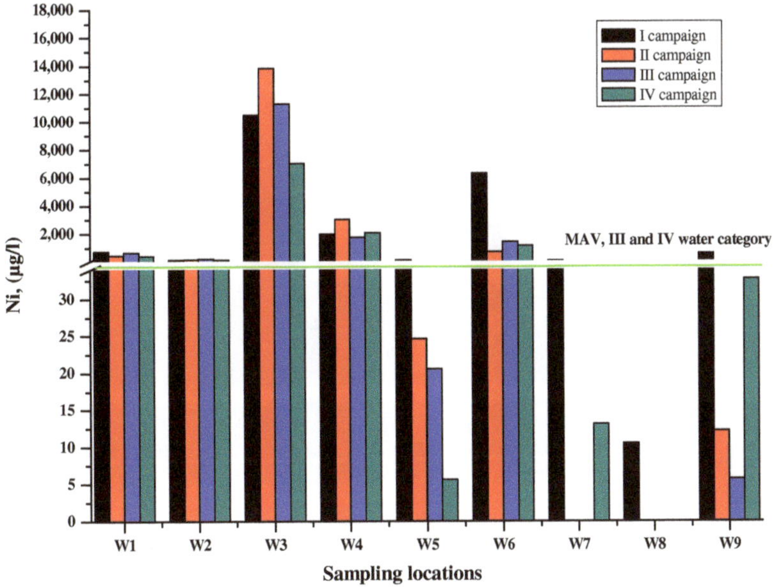


**Figure 5.** Concentration of Ni ions in the realized sampling campaigns oat defined locations.

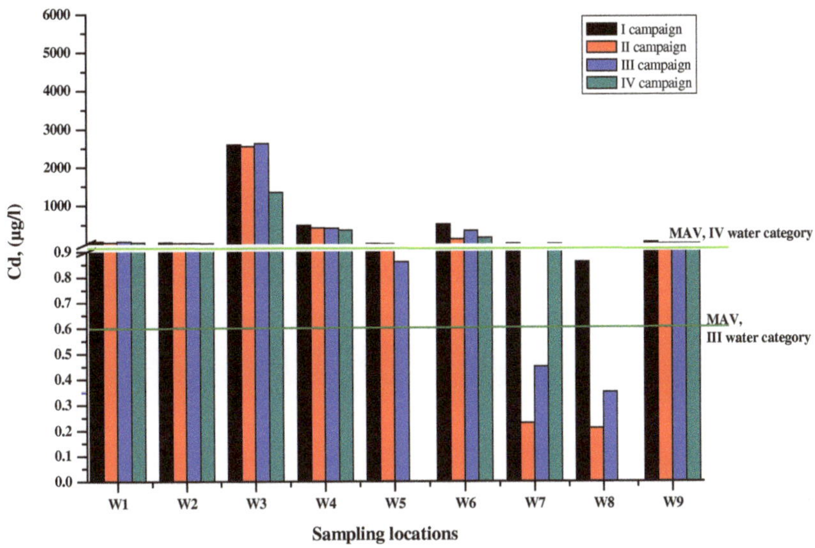


**Figure 6.** Concentration of Cd ions in the realized sampling campaigns at defined locations.

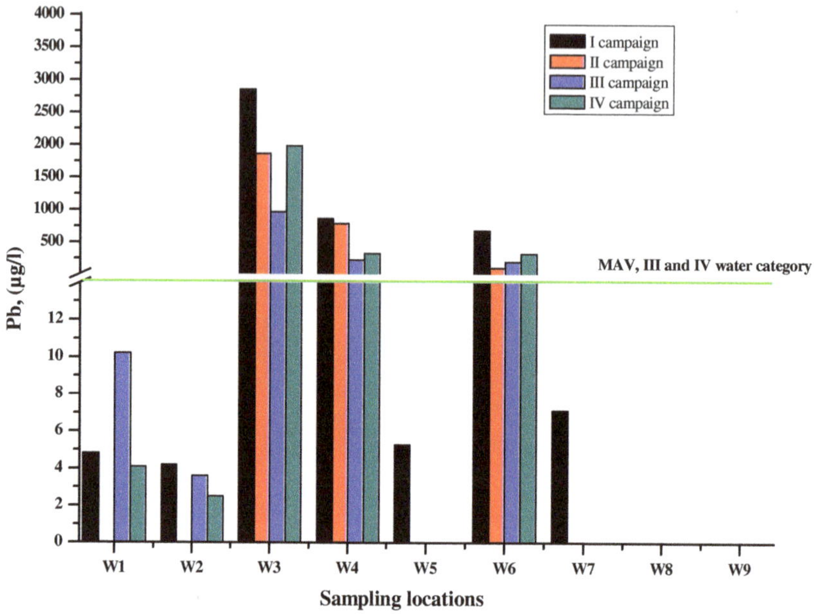


**Figure 7.** Concentration of Pb ions in the realized sampling campaigns at defined locations.

The calculated average and SD values (plus direction of error bars) for Ni, Cd and Pb are presented in Figure 8.

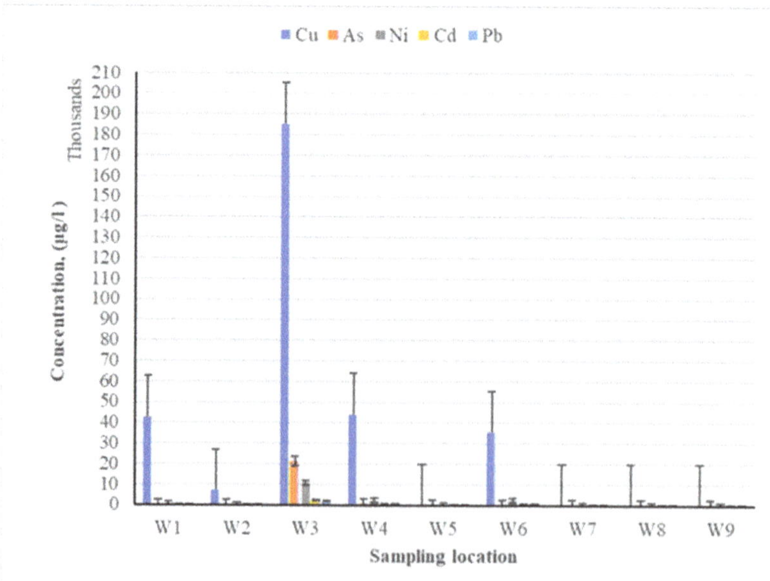


**Figure 8.** The average and SD values for concentration of Cu, As, Ni, Cd and Pb ions.

During the sampling campaigns, the impact of AMD water from overburden (W1), AMD water from flotation tailings (W2), and metallurgical wastewater (W3) on the surface water of the Bor River (W4) was investigated. The results for the concentration of Cu,

As, Ni, Cd, and Pb ions in all three kinds of wastewater that inflow into the Bor River have shown the following: the concentration of Cu ions in water samples, obtained by the natural leaching of overburden (W1), is in the range of 31.6–57.2 mg/L, average ± SD (mg/L) values, 42.68 ± 12.30. The concentration of Cu ions in the AMD water from flotation tailings ranges up to 10 mg/L, average ± SD (mg/L) values, 7.10 ± 2.31, while the concentration of Cu ions in the metallurgical wastewater ranges from 155.27 to 207.9 mg/L, average ± SD (mg/L) values, 185.25 ± 26.05. The obtained results, presented in Figures 3 and 8, have confirmed that the concentration of Cu ions had the highest values in metallurgical wastewater (W3). The literature data have also confirmed that Cu ions are also detected in a similar type of mining-influenced water [24]. The concentration of Cu ions in AMD water at Parys Mountain (Wales) was in the range of 7–44 mg/L. Similar values are registered in the other copper mines in this area [27].

Based on the values for standard deviation, the possibility that the values could be found outside the range is lower for samples from location W1, W2 and W3 (Figure 8) compared to samples from other investigated locations.

The concentration of arsenic ions in the metallurgical wastewater is extremely high (more than 19,000 μg/L) in all samples, average ± SD (μg/L) values, 21,445.97 ± 2725.33. As the metallurgical wastewater inflows into the Bor River, it is clear that a high concentration of As ions in the Bor River is the result of their presence in W3. Also, a high concentration of Ni, Cd, and Pb ions is the result of the presence of those ions in W3, which inflow in W4. The literature data confirm that arsenic is a toxic and harmful impurity in the copper production process. The characteristic of As ions is that they can be found in metallurgical wastewater, and they could have a negative impact on the environment and biodiversity [28,29]. The results have confirmed that metallurgical wastewater (W3) is one of the main pollutants of the Bor River. Based on the results that the concentration of As ions is mostly in the range of SD values, it is real to expect that the concentration of As ions will be in this range in cases of unchanged mining and metallurgy activities (Figure 8).

The results of the concentration of Cu, As, Ni, Cd, and Pb ions in the Bor River (sample location W4), presented in Figures 3–7, have approved that each value is higher than the MAV:

- Cu ions: The highest concentration of Cu ions was registered during the IV sampling campaign, and it was approximately 63 times higher than the MAV. The minimum value was registered during the III sampling campaign, and it was almost 25 times higher than the MAV.
- As ions: The highest concentration of As ions (2552.91 μg/L) was about 26 times higher than the MAV, and was registered during the I sampling campaign. The minimum value (74.4 μg/L) was registered during the III sampling campaign, and it was lower than the MAV.
- Ni ions: The highest concentration of 3020.6 μg/L was registered during the II sampling campaign, and it was about 90 times higher than the MAV. The minimum value (1740.59 μg/L) was registered during the III sampling campaign, and it was 50 times higher than the MAV.
- Cd ions: The highest concentration of 480.15 μg/L was registered during the I sampling campaign, and it was about 533 times higher than the MAV. The minimum value (355.7 μg/L) was registered during the IV sampling campaign, and it was approximately 400 times higher than the MAV.
- Pb ions: The highest concentration of 867.60 μg/L was registered during the I sampling campaign, and it was about 62 times higher than the MAV. The minimum value (231.52 μg/L) was registered during the III sampling campaign, and it was approximately 16 times higher than the MAV.

Generally, the results of samples from the Bor River (W4) have shown that only the concentration of As ions was lower than the MAV in only one sampling campaign. The concentration of all other heavy metal ions was above the MAV, even 533 times higher in the case of Cd ions.

The (W5) concentration of Cu, As, and Pb ions in all cases was lower than the MAV in the Krivelj River. The concentrations during the, I, and II sampling campaign were increased by Cd ions, and the concentration of Ni ions was increased only in the I sampling campaign.

The characteristic of the Ravna River (W7) is that the concentrations of Cu, As, and Pb ions during each sampling campaign had lower values than the MAV, while the concentrations of Ni and Cd ions had higher values than the MAV during the I and IV sampling campaigns. During the I sampling campaign, the concentrations of both elements were increased, and during the IV sampling campaign, only the concentration of Cd ions was increased.

The characteristic of the Bela River (W6), which is formed by merging the Bor, Krivelj, and Ravna rivers, was that the concentrations of Cu, As, Ni, Cd, and Pb ions during each sampling campaign were higher than the MAV. By comparison with the values of appropriate elements in the rivers that form the Bela River, it is clear that the Bor River is a carrier of heavy metal ions that cause an enormous amount of pollution downstream towards the Danube River.

The characteristic of the Timok River (W8), which flows through the area out of impact of the mining and metallurgy activities from the study area, is that the concentration of all monitored element ions is lower than the MAV.

No negative impact of the mining and metallurgical activities on water in the Veliki Timok River (W9) is recorded.

### 3.1.2. pH Values

The pH values, measured during the realized sampling campaigns, are shown in Figure 9.

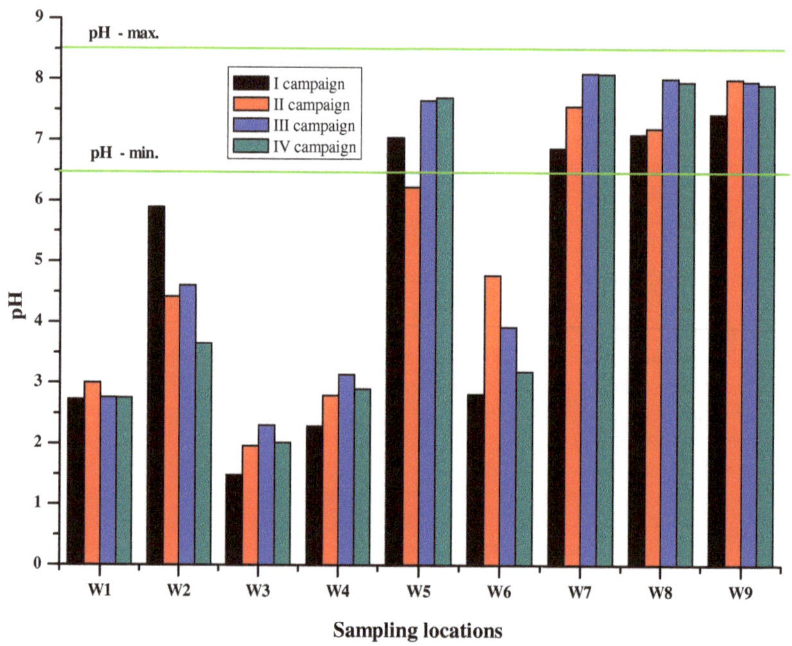


**Figure 9.** pH values at different sampling locations for each sampling campaign.

It is evident from Figure 9 that mining-affected water (W1 and W2) and metallurgical wastewater (W3) had increased acidity, which was also shown in the literature data for similar water types [30]. Low pH values, also a characteristic of the Bor River, were at the minimum in the I sampling campaign (2.29). In the same campaign, the pH value was the lowest at the W3 location (1.49). The above data have confirmed once again a dominant impact of metallurgical wastewater on the Bor River characteristics, but also that the present impact of mining has affected water.

The Bela River (W6), formed by merging the Bor, Krivelj, and Ravna rivers, is also characterized by low pH values compared to the pH values of surface water from locations W5, W7, W8, and W9. The Krivelj River (W5) is characterized by a pH in the range of 6.23–7.7, but after merging the Bor River with a pH value in the range 2.29–3.14, the pH value of the newly formed river, the Bela River (W6), ranges between 2.82 and 3.78.

The pH values of surface water from location W7, W8 and W9 are within the MAV range (6.5–8.5).

### 3.1.3. Electrical Conductivity

One of the water properties is its electrical conductivity, i.e., the ability of water to conduct electricity. The electrical conductivity of water shows how many dissolved substances, salts, minerals, and chemicals are present in it. Salts and various chemicals, dissolved in water, consist of positively and negatively charged ions that conduct electricity in water. The increased electrical conductivity of water indicates its possible pollution. The literature data have confirmed the elevated electrical conductivities for the same water types [31]. Figure 10 presents the measured values of electrical conductivity during the realized sampling campaigns.

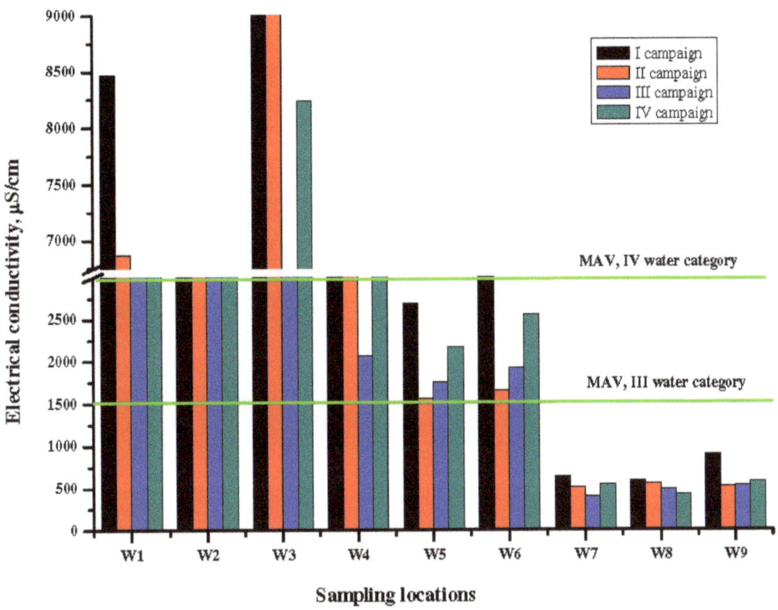


**Figure 10.** Electrical conductivity at different sampling locations for each sampling campaign.

The electrical conductivity values for wastewater from locations W1, W2, and W3 were very high, ranging from above 3500 µS/cm to 8500 µS/cm, indicating the presence of salts and various chemicals dissolved in water and whose ions enable the conduction of electricity. This confirms the presence of high concentrations of Cu, As, Ni, Cd, and Pb ions in the investigated samples, as shown in Figures 3–7. It was not possible to measure the electrical conductivity of water from the W3 samples in two sampling campaigns (values were above the detection limit of an instrument, 19,990 µS/cm), confirming very high concentrations of metal ions in this water.

The surface water, under a direct impact of mining and metallurgical activities (W4, W5, and W6), is characterized by slightly lower values of electrical conductivity. However, they are still higher than 1500 μS/cm, which is the upper limit for class III water according to the Serbian legislation. The characteristic of surface water from locations W7 and W8 is that the values are below 1500 μS/cm [25,26].

The electrical conductivity of the Veliki Timok River (W9), during certain sampling campaigns, was higher than those values of water samples from locations W7 and W8. This has confirmed the impact of the Bela River (W5) on the electrical conductivity of the Veliki Timok River (W9) after merging the Timok River (W8) and Bela River (W5). A similar situation is registered in the area of the Majdanpek copper mine [32].

### *3.2. Characteristics of the Bor River and Bela River (Period 2022–2024)*

During 2022, the concentrations of Cu, As, Ni, and Pb ions, pH value, and electrical conductivity in the Bor River (W4) were monitored by an authorized institution for the needs of the city of Bor. Also, the monitoring of the Bela River (W6) was started in 2023. The available reports were downloaded from the website of the city of Bor and used for a comparison with the maximum values obtained in four sampling campaigns (September 2020–June 2021). Table 3 presents the data for the pH, electrical conductivity and concentration of Cu ions. Based on the fact that the concentrations of As, Ni, Cd, and Pb ions were expressed in the same measurement units, the results are presented by diagrams in Figure 11.

**Table 3.** pH values, EC and concentration of Cu ions, adapted from Refs. [33–37].

| Parameters | Sampling Location/Date | | | | | | | | | |
|---|---|---|---|---|---|---|---|---|---|---|
| | W4/ 20–21 | W6/ 20–21 | W4/ 07.22 | W4/ 10.22 | W4/ 10.23 | W6/ 10.23 | W4/ 11.23 | W6/ 11.23 | W4/ 03.24 | W6/ 03.24 |
| pH | 2.29 | 2.82 | 8.00 | 7.90 | 7.37 | 7.5 | 7.26 | 7.48 | 7.84 | 7.91 |
| EC, (μS/cm) | 5896 | 3996 | 2180 | 1803 | 1699 | 1557 | 1467 | 1152 | 1714 | 1232 |
| Cu, (mg/L) | 62.9 | 60.45 | 1.58 | 0.891 | 0.031 | 0.028 | 0.037 | 0.012 | 0.03 | 0.022 |

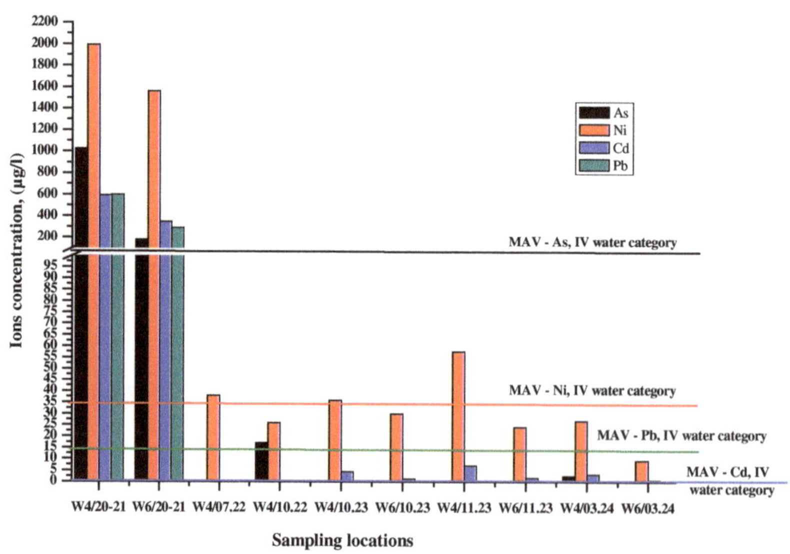


**Figure 11.** Comparison of concentration of As, Ni, Cd and Pb ions in different periods (2020–2024).

Changes made by the Serbia Zijin Copper d.o.o Bor had a goal to stop the discharge of mining-influenced water (locations W1 and W2) and metallurgy wastewater (locations W3) into the Bor River (W4), reuse this water and reduce the impact on the human and natural environments around the mining area as well as at the other locations [38,39]. The metallurgical wastewater (W3) treatment in a newly built facility, pumped out and neutralized water from location W2, and redirection of the W1 flow into the active flotation tailing dump led to a drastic decrease in the concentrations of Cu, As, Ni and Pb ions in the Bor River (W4) (Table 3). The concentration of Cu ions, determined in July 2022, had a maximum value of 1.58 mg/L, which is almost 40 times less than the maximum concentration of Cu ions determined in the IV sampling campaign (62.9 mg/L, Figure 3). During the last measured time (18 March 2024), the concentration of Cu ions in the Bor River was 0.03 mg/L. It is clear that the concentration of Cu ions in the Bor River in March 2024 is about 50 times lower than the concentration in July 2022.

As a consequence of decreasing the concentration of the Cu ion concentration in the Bor River, the concentration of Cu ions in the Bela River (W6) is also decreased (Table 3). The maximum value for the concentration of Cu ions in the Bela River (60.45 mg/L) is registered in the I sampling campaign (Figure 3), and it is about 2200 times higher than the value registered in October 2023. The same concentration value of Cu ions in the Bela River is registered in March 2024.

The results of the concentration of Cu ions in the Bor River have confirmed that the values are below the MAV for the III water class in the period 2023–2024 (Table 2).

The concentration values of Ni and Cd ions were also decreased in samples from the Bor and Bela rivers, but some values were higher than the MAV.

The concentration of arsenic ions in some cases was below the detection limit of the applied analytical method, and the values were below the MAV for two samples from the Bor River.

The concentration of lead ions was below the detection limit of the applied analytical method for both rivers.

Measurements of the pH value have confirmed that the values were in the MAV range for the IV surface water class according to the data from Table 2. Also, the values of electrical conductivity decreased. It was confirmed that the concentration of metal ions decreased. The values for the pH and electrical conductivity for all samples from the Bor and Bela rivers were in the range of MAVs for the monitored period.

The concentration of 57.5 μg/L for the Ni ions (the highest value in the period 2022–2024) was almost 50 times lower than the highest concentration of 3020.6 μg/L, registered during the II sampling campaign in the period 2020–2021 (Figure 5), but still above the MAV value.

The concentration of Cd ions in the Bor River is still higher than the MAV, but the values in samples from the Bela River are lower than the MAV in two samples (sampling in October 2023 and March 2024).

The characteristics of both rivers (the Bor and Bela rivers), sampled in the period 2022–2024, are much better compared to the characteristics from the period September 2020–June 2021. The study, realized in the earlier period, has confirmed that very serious environmental impacts are detected 30 km along the Bela River. Inflowing of the untreated AMD water from an overburden disposal as well as from flotation tailing dumps will continue to significantly impact the local and regional water resources [40].

## 4. Conclusions

The obtained values for concentrations of Cu, As, Ni, Cd, and Pb ions in the AMD water from overburden (W1), flotation tailing (W2), and metallurgical wastewater (W3) that flow into the Bor River have confirmed that the concentration of Cu ions has the highest values in the metallurgical wastewater samples. The concentration of arsenic ions in the metallurgical wastewater that inflow into the Bor River was extremely high as well as the concentration of As ions in the Bor River. Also, the high concentration of Ni, Cd, and Pb ions in the Bor River is a consequence of the presence of those ions in the metallurgical wastewater (W3). These results have confirmed that metallurgical wastewater (W3) from on-going mining activity is one of the main pollutants of the Bor River. At the same time, the AMDs that inflow into the Bor River from different locations are the carriers of heavy metal ions and directly affect the quality of the Bor River.

The concentration of Cu, As, and Pb ions in the Krivelj River (W5) was in all cases lower than the MAV.

The characteristic of Ravna River (W7) is that the concentrations of Cu, As, and Pb ions, in each sampling campaign, had lower values than the MAV.

Cd and Ni ions had increased concentrations during some sampling campaigns in the Krivelj (W5) and Ravna rivers (W7).

The concentrations of Cu, As, Ni, Cd, and Pb ions in the Bela River (W6), which is formed by merging the Bor, Krivelj, and Ravna rivers, were higher than the MAV in each sampling campaign.

The characteristic of the Timok River (W8), which flows through the area that is out of impact of the mining and metallurgy activities from the study area, is that the concentration of each monitored element ion is lower than the MAV.

The negative impact of the mining and on-going mining activities on water in the Veliki Timok River (W9) was also recorded.

The pH values during each sampling campaign were low for the AMD water from different locations, metallurgical wastewater, Bor River, and Bela River.

The electrical conductivity of metallurgical wastewater in almost all sampling campaigns had a value which was higher than the instrument detection limit.

The concentration of monitored elements in the Bor River and Bela River after stopping the inflowing of metallurgical wastewater and AMD water from overburden and flotation tailing dumps into the Bor River (detected in the period 2022–2024) confirmed that the quality of the mentioned rivers is better in comparison with the quality of those rivers in the period September 2020–June 2021. All activities realized by a company that manages the mining and on-going mining activities, within the investigated study area, had a positive effect on the environment.

This research can represent a reference point for future studies in eastern Serbia or other areas where the copper overburden and flotation tailings have been deposited, using the observed concentrations of metal ions. Also, further activities, aimed at preventing the inflow of untreated mining-influenced water and industrial wastewater into the surrounding watercourses, will lead to economic and ecological effects.

**Author Contributions:** Individual contributions of authors are as follows: conceptualization, R.M. and Z.Š.; methodology, R.M.; software, Z.Š. and V.M.M.; validation, V.G. and J.P.; formal analysis, R.M. and R.K.; investigation, R.M., V.G., R.K. and J.P.; resources, R.M., V.G. and J.P.; data curation, V.G., R.K. and V.M.M.; writing—original draft preparation R.M. and V.M.M.; writing—review and editing, R.M., V.M.M., Z.S. and Z.Š.; visualization, B.F. and Z.Š.; supervision R.M. and B.F.; project administration V.M.M. All authors have read and agreed to the published version of the manuscript.

**Funding:** The research presented in this paper was done with support of the Ministry of Science, Technological Development and Innovation of the Republic of Serbia, within the scientific research work at the Mining and Metallurgy Institute Bor, (Contract No. 451-03-66/2024-03/200052) and University of Belgrade, Technical Faculty in Bor (Contract No. 451-03-65/2024-03/200131).

**Data Availability Statement:** The original contributions presented in the study are included in the article, further inquiries can be directed to the corresponding author.

**Conflicts of Interest:** The authors declare no conflicts of interest.

## References


1. Helios Rybicka, E. Impact of Mining and Metallurgical Industries on the Environment in Poland. *Appl. Geochem.* **1996**, *11*, 3–9. [CrossRef]
2. Wolkersdorfer, C.; Mugova, E. Effects of Mining on Surface Water. In *Human Pressures and Management of Inland Waters, Encyclopedia of Inland Waters*, 2nd ed.; Irvine, K., Chapman, D., Warner, S., Eds.; Elsevier Inc.: Amsterdam, The Netherlands, 2022; pp. 170–190. [CrossRef]
3. Petronijević, N.; Radovanović, D.; Stanković, S.; Sokić, M.; Kamberović, Ž.; Onjia, A.; Jovanović, G. Stability of Solid Residue After Integral Treatment of Acid Mine Drainage. *Metall. Mater. Data* **2023**, *1*, 57–63. [CrossRef]
4. Adnan, M.; Xiao, B.; Xiao, P.; Zhao, P.; Li, R.; Bibi, S. Research Progress on Heavy Metals Pollution in the Soil of Smelting Sites in China. *Toxics* **2022**, *10*, 231. [CrossRef] [PubMed]
5. Isvoran, A.; Roman, D.L.; Dascalu, D.; Vlad-Oros, B.; Ciorsac, A.; Pitulice, L.; Jonovic, R.; Stevanovic, Z.; Ostafe, V. Human Health Effects of Heavy Metal Pollution in the Cross-Border Area of Romania and Serbia: A Review. *Ecol. Chem. Eng. S* **2021**, *28*, 365–388. [CrossRef]
6. Kukusamude, C.; Sricharoen, P.; Limchoowong, N.; Kongsri, S. Heavy Metals and Probabilistic Risk Assessment via Rice Consumption in Thailand. *Food Chem.* **2021**, *334*, 127402. [CrossRef] [PubMed]
7. Freeland-Graves, J.H.; Sachdev, P.K.; Binderberger, A.Z.; Sosanya, M.E. Global Diversity of Dietary Intakes and Standards for Zinc, Iron, and Copper. *J. Trace Elem. Med. Biol.* **2020**, *61*, 126515. [CrossRef] [PubMed]
8. Ekeanyanwu, C.L.; Alisi, C.S.; Ekeanyanwu, R.C. Levels of Aflatoxin M1 and Selected Heavy Metals (Pb, Cd, Cr, Cu, Zn, Fe, As, and Hg) in the Breast Milk of Lactating Mothers in South Eastern, Nigeria. *Food Control* **2020**, *112*, 107150. [CrossRef]
9. Palma-Lara, I.; Martínez-Castillo, M.; Quintana-Perez, J.C.; Arellano-Mendoza, M.G.; Tamay-Cach, F.; Valenzuela-Limon, O.L.; García-Montalvo, E.A.; Hernandez-Zavala, A. Arsenic Exposure: A Public Health Problem Leading to Several Cancers. *Regul. Toxicol. Pharmacol.* **2020**, *110*, 104539. [CrossRef] [PubMed]
10. Qin, S.; Liu, H.; Nie, Z.; Rengel, Z.; Gao, W.; Li, C.; Zhao, P. Toxicity of Cadmium and its Competition with Mineral Nutrients for Uptake by Plants: A Review. *Pedosphere* **2020**, *30*, 168–180. [CrossRef]
11. Buha, A.; Jugdaohsingh, R.; Matovic, V.; Bulat, Z.; Antonijevic, B.; Kerns, J.G.; Goodship, A.; Hart, A.; Powell, J.J. Bone Mineral Health is Sensitively Related to Environmental Cadmium Exposure-Experimental and Human Data. *Environ. Res.* **2019**, *176*, 108539. [CrossRef] [PubMed]
12. Dorea, J.G. Environmental Exposure to Low-Level Lead (Pb) Co-Occurring with other Neurotoxicants in Early Life and Neurodevelopment of Children. *Environ. Res.* **2019**, *177*, 108641. [CrossRef]
13. Available online: www.ekologija.gov.rs/sites/default/files/2022-03/program_upravljanja_otpadom_eng_-_adopted_version.pdf (accessed on 12 April 2024).
14. Available online: https://www.stat.gov.rs/en-us/publikacije/publication/?p=13253 (accessed on 12 March 2024).
15. Shilpa, S.S.; Deepthi, D.; Harshitha, S.; Shipra, S.; Prashanth, B.N.; Suchetha, K.N.; Harishkumar, M. Environmental Pollutants and their Effects on Human Health. *Heliyon* **2023**, *9*, e19496. [CrossRef] [PubMed]
16. Đorđievski, S.; Ishiyama, D.; Ogawa, Y.; Stevanović, Z. Mobility and Natural Attenuation of Metals and Arsenic in Acidic Waters of the Drainage System of Timok River from Bor Copper Mines (Serbia) to Danube River. *Environ. Sci. Pollut. Res.* **2018**, *25*, 25005–25019. [CrossRef] [PubMed]
17. Spooren, J.; Binnemans, K.; Björkmalm, J.; Breemersch, K.; Dams, J.; Folens, K.; González-Moya, M.; Horckmans, L.; Komnitsas, K.; Kurylak, W.; et al. Near-Zero-Waste Processing of Low-Grade, Complex Primary Ores and Secondary Raw Materials in Europe: Technology Development Trends. *Resour. Conserv. Recycl.* **2020**, *160*, 104919. [CrossRef]
18. Štirbanović, Z.; Gardić, V.; Stanujkić, D.; Marković, R.; Sokolović, J.; Stevanović, Z. Comparative MCDM Analysis for AMD Treatment Method Selection. *Water Resour. Manag.* **2021**, *35*, 3737–3753. [CrossRef]
19. Filimon, M.N.; Popescu, R.; Horhat, F.G.; Voia, O.S. Environmental Impact of Mining Activity in Bor Area as Indicated by the Distribution of Heavy Metals and Bacterial Population Dynamics in Sediment. *Knowl. Manag. Aquat. Ecosyst.* **2016**, *417*, 30. [CrossRef]
20. Stevanović, Z.; Kovačević, R.; Marković, R.; Gardić, V.; Vulpe, C.B.; Boros, B.; Menghiu, G. State of the Surface Waters in Cross Border Region of Eastern Serbia and Caras Severin County—Moldova Noua in Romania. *Stud. UBB Chem.* **2021**, *66*, 309–328. [CrossRef]
21. Panias, D. Consequences of Environmental Issues on Sustainability of Metal Industries in Europe: The Case Study of Bor. *Metalurgija* **2006**, *054*, 239–250.
22. Petković, S. The Traces of Roman Metallurgy in Eastern Serbia. *J. Min. Metall. Sect. B Metall.* **2009**, *45*, 187–196. [CrossRef]
23. Al-Robai, S.A. Ecological Risk Evaluation of Heavy Metals in Soils near a Water Dam in Baljurashi, KSA, and Their Accumulation in Dodonaea viscosa. *Sustainability* **2023**, *15*, 15646. [CrossRef]
24. Hadži Jordanov, S.; Maletić, M.; Dimitrov, A.; Slavkov, D.; Paunović, P. Waste Waters from Copper Ores Mining/Flotation in "Bučim" Mine: Characterization and Remediation. *Desalination* **2007**, *213*, 65–71. [CrossRef]
25. Regulation on the Categorization of Watercourses. ("Official Gazette of SRS", No. 5/1968-61).
26. Regulation on Limit Values of Pollutants in Surface and Underground Waters and Sediment and Deadline for Reaching Them. ("Official Gazette of RS", No. 50/2012).
27. Available online: https://www.imwa.info/docs/imwa_2004/IMWA2004_27_Mullinger.pdf (accessed on 15 April 2024).

28. Perez, K.; Toro, N.; Galvez, E.; Robles, P.; Wilson, R.; Navarra, A. Environmental, Economic and Technological Factors Affecting Chilean Copper Smelters—A Critical Review. *J. Mater. Res. Technol.* **2021**, *15*, 213–225. [CrossRef]
29. Houngaloune, S.; Hiroyoshi, N.; Ito, M. Development of Arsenic Stabilization Techniques in the Mining and Metallurgical Industries—A Review. In Proceedings of the MMIJ Spring Conference (Resources and Materials 2013), Sapporo, Japan, 3–5 September 2013; pp. 297–300.
30. Nordstrom, D.K.; Blowes, D.W.; Ptacek, C.J. Hydrogeochemistry and Microbiology of Mine Drainage: An Update. *Appl. Geochem.* **2015**, *57*, 3–16. [CrossRef]
31. Šerbula, S.; Stanković, V.; Živković, D.; Kamberović, Ž.; Gorgievski, M.; Kalinović, T. Characteristics of Wastewater Streams Within the Bor Copper Mine and Their Influence on Pollution of the Timok River, Serbia. *Mine Water Environ.* **2016**, *35*, 480–485. [CrossRef]
32. Atlagić, J.Č.; Andjus, S.; Đuknić, J.; Milivojević, A.; Kulić, K.; Pokrajac, J.; Tubić, B.; Paunović, M. Assessment of the Impact of Copper Mining and Related Industrial Activities on Aquatic Macroinvertebrate Communities of the Pek River (Serbia). *Arch. Biol. Sci.* **2021**, *73*, 291–301. [CrossRef]
33. Available online: https://bor.rs/wp-content/uploads/2022/08/izvestaj.pdf (accessed on 12 April 2024).
34. Available online: https://bor.rs/wp-content/uploads/2022/11/Izvestaj-07.11.2022.pdf (accessed on 12 April 2024).
35. Available online: https://bor.rs/wp-content/uploads/2023/11/4117-23-GU-Bor-vode-09.10.2023.-15.10.2023.-Izvestaj.pdf (accessed on 12 April 2024).
36. Available online: https://bor.rs/wp-content/uploads/2023/11/4771-23-GU-Bor-vode-12.11.2023.-Izvestaj.pdf (accessed on 12 April 2024).
37. Available online: https://bor.rs/wp-content/uploads/2024/03/1067-24-GU-Bor-vode-11.03.2024.pdf (accessed on 12 April 2024).
38. Abiye, T.A.; Ali, K.A. Potential Role of Acid Mine Drainage Management Towards Achieving Sustainable Development in the Johannesburg Region, South Africa. *Groundw. Sustain. Dev.* **2022**, *19*, 100839. [CrossRef]
39. Baloyi, J.; Ramdhani, N.; Mbhele, R.; Ramutshatsha-Makhwedzha, D. Recent Progress on Acid Mine Drainage Technological Trends in South Africa: Prevention, Treatment, and Resource Recovery. *Water* **2023**, *15*, 3453. [CrossRef]
40. Markovic, R.; Bessho, M.; Masuda, N.; Stevanovic, Z.; Bozic, D.; Apostolovski Trujic, T.; Gardic, V. New Approach of Metals Removal from Acid Mine Drainage. *Appl. Sci.* **2020**, *10*, 5925. [CrossRef]

*metals*

MDPI

*Article*

# Electrochemical Behaviour of an Au-Ge Alloy in an Artificial Saliva and Sweat Solution

**Gyöngyi Vastag [1], Peter Majerič [2,3], Vojkan Lazić [4] and Rebeka Rudolf [2,3,*]**

1 Faculty of Sciences, University of Novi Sad, 21000 Novi Sad, Serbia
2 Faculty of Mechanical Engineering, University of Maribor, 2000 Maribor, Slovenia; peter.majeric@um.si
3 Zlatarna Celje d.o.o., 3000 Celje, Slovenia
4 School of Dental Medicine, University of Belgrade, 11158 Belgrade, Serbia; vojkan.lazic@stomf.bg.ac.rs
* Correspondence: rebeka.rudolf@um.si

**Abstract:** In modern times, more and more different materials (including alloys) are in direct contact with human electrolytes (sweat, saliva, lymph, blood, etc.). One of the most important properties for the use of these materials is therefore their chemical inertness or resistance to corrosion when they are in contact with human electrolytes. Consequently, during the development of such new materials, it is necessary to study and understand their basic electrochemical behaviour in a given environment. The purpose of this research was to monitor the electrochemical behaviour of the new Au-Ge alloy in artificial sweat and artificial saliva solutions, depending on the electrolyte composition and exposure time. This new alloy represents a potential material for use in dentistry or for jewellery. The obtained results of the study show that the immersion time and the pH value have a significantly greater influence on the corrosion resistance of the new Au-Ge alloy than the composition of the electrolyte solution. The results of the SEM/EDX analysis additionally confirm the main results of the electrochemical measurements.

**Keywords:** Au-Ge alloy; electrochemical properties; characterisation

## 1. Introduction

Nowadays, metals and alloys, apart from having significant and well-known applications as structural materials, increasingly find their role in our daily life as coins, zippers, buttons, in the form of various fashion accessories (jewellery, piercings, eyeglass frames, wrist watches, etc.), as a different type of orthopaedic and dental amalgam, implants, or prostheses, etc. All the above bring metals and alloys into short- or long-term contact with various physiological fluids (electrolytes), like sweat, saliva, lymph, blood, etc. Contacts between alloys and different types of human electrolytes can lead to electrochemical reactions (electrochemical corrosion). These electrochemical reactions are very important, because, on the one hand, they can change the metals' (alloys) features, and they could also have an impact on human tissue characteristics and properties [1,2]. Because of this, one of the most important contacts for the usage of new materials is their chemical inertness or corrosion resistance during contact with human electrolytes. Consequently, during the development of new materials (alloys), it is essential to examine and understand their basic electrochemical behaviour in the given environment.

For a long time, nickel-based alloys have been used widely for the above purposes, but it is well known that nickel ions, which are formed through electrochemical corrosion under the influence of human sweat, can cause a strong allergic reaction in a high percentage of the population [3,4]. Consequently, many studies have been conducted so far with the aim of achieving an economically affordable and sustainable alternative to the conventional nickel alloy. Developing gold alloys for fashion accessories without nickel presents a challenge, as nickel provides the strength and hardness of the resulting alloy and has favourable casting characteristics for lost-wax casting, providing ease of flow and useful shrinkage during solidification, as well as good mechanical properties for rolling and machining during production. It is also a whitening element, suitable for producing commercially more interesting white gold products, which have increased in popularity on the jewellery market in the last few decades. Alternatives for nickel include palladium, platinum, or titanium, which increase the price of the alloy or the technological difficulty of production.

An alternative alloying element was found in Germanium during an attempt to produce a gold alloy with characteristics suitable for production, such as with the case of using nickel [5]. The design of the new Au-Ge alloy is aimed at achieving properties that are important for use in dentistry as a carrier alloy for porcelain dental reconstructions, and for use in jewellery for the production of a white Au pre-alloy. In both cases, the new Au-Ge alloy will need to have good mechanical characteristics and high biocompatibility with corrosion resistance. In jewellery, most manufacturers of white Au alloys use Ni as a key element for the final colouring, and some also use Pd. In the case of using Ni, the key issue is the release of nickel ions, which cause skin allergies, such as the Au-Ge alloy, which is said to have high electrical and thermal conductivity and corrosion resistance due to the Au content. Using Ge decreases the hardness, increases the flowability of the alloy, decreases shrinkage during alloy casting, and is more economically favourable than using other precious metals. This alloying element is a potentially appropriate component in the conventional gold–copper–zinc alloys for established production techniques in jewellery manufacturing companies. High amounts of germanium produce brittleness and decrease hardness in gold alloys, which is not practical for mechanical treatment. As such, a gold alloy containing a small amount of Ge, labelled as an Au-Ge alloy, was produced to investigate its properties as a substitute for nickel. On the other hand, the crucial information is that Ge compounds are relatively less toxic when compared to other metalloids and metals [6]. Based on this, Ge is used for medical purposes as different agents to repair bones with infected bone defects [7]. Ge is not cytotoxic at a low doping level and is not carcinogenic [8], and even appears to inhibit inflammatory processes in the effective clinical treatment for mastitis and other inflammatory diseases [9,10]. He has been considered biocompatible, and, in the antimicrobial surface coatings, has been recommended as a bio-implant material for the first time [11]. The aim of this research was to investigate the electrochemical behaviour of one new Au-Ge alloy as potential jewellery or dental materials in an artificial sweat and artificial saliva solution. Human sweat has a complex composition (different types of electrolytes, organic acids, carbohydrates, amino acids, etc.) [12,13], which varies greatly in the function of different physical, environmental, and pharmacological conditions, such as age and sex, body weight, general body condition, etc. Human saliva, like sweat, can vary a lot in its composition, also depending on the age and gender of the patient, but oral hygiene and eating habits, such as the time of day, often have an additional effect [14,15]. Because of that, different types of artificial saliva are usually used in investigations [16]. The most used artificial saliva for the investigation of the electrochemical properties of dental metals and alloys is a modified Fusayama's solution [17].

## 2. Materials and Methods

The Au-Ge alloy was produced with raw materials with a high purity as follows: Au 99.99%, Cu 99.99%, Zn 99.99%, Ge 99.99%, and Ir 99.99% (Legor Group S.p.A., Bressanvido, Italy), with a proprietary final composition [5]. The casting of the alloy was performed in a protective atmosphere of Ar 5.0 with a clay graphite crucible at 1100 °C. The casting was performed with a rod with a diameter of 20 mm, which was then rolled through steps of deformation into a square profile strip with a thickness of 1 mm by 10 mm width. The strip was then cut into tiles with a length of 10 mm each. The final tiles, with the dimensions 10 × 10 × 1 mm, were used as samples for the artificial saliva and sweat solution testing.

The artificial sweat composition used was as follows: 20 g/L NaCl, 17.5 g/L, $NH_4Cl$, 5 g/L acetic acid, and 15 g/L lactic acid. The artificial sweat solution pH value was adjusted to 4.7 by NaOH [18,19].

The used artificial saliva (modified Fusayama's solution) composition was as follows: 0.400 g/L NaCl, 0.400 g/L KCl, 0.795 g/L $CaCl_2 \cdot 2H_2O$, 0.690 g/L $NaH_2PO_4 \cdot H_2O$, 0.005 g/L $Na_2S \cdot 9H_2O$, and 1.000 g/L urea. The original pH of the solution was 4.81. In human saliva, there is often a short-term decrease in the pH value that occurs during the intake of acidic foods and drinks, as well as due to the secretion of stomach acid. In order to obtain information about the effect of lowering the pH value on the electrochemical behaviour of the investigated alloy, in addition to the original artificial saliva solution, research was also conducted on the artificial saliva solution in which the pH was adjusted to 2.51 using lactic acid [14].

A three-electrode cell was used for the electrochemical measurements, with the Au-Ge alloy as the working electrode, the saturated calomel electrode (SCE) as the reference electrode, and the platinum as the counter electrode. The working electrode was constructed from an Au-Ge alloy plate, with an exposed area of 0.5 $cm^2$, embedded in an epoxy resin. The electrode was wet polished with SiC papers (grit sizes of 800 and 1200), rinsed with acetone and double distilled water, and then immersed into the electrolyte solution. A PC controlled VoltaLab PGZ 301 (Radiometer Analytical SAS, Villeurbanne, France) was applied for electrochemical measurements. The potentiodynamic measurements were performed after immersion times of 1 h, 24 h, 72 h, and after 7 days at room temperature (approximately 25 °C). The potential was scanned between OCP and −700 mV/SCE in the cathodic direction and back to the anodic direction (+700 mV/SCE) at a scan rate of 1 mV/s.

The measurements of electrochemical impedance spectra (EIS) were also performed after immersion times of 1 h, 24 h, 72 h, and after 7 days at the open-circuit potential, (OCP) and at room temperature. The EIS measurements were carried out over a frequency range of 0.01 Hz–10 kHz, using 10 mV amplitude of sinusoidal voltage.

An additional scanning electron microscopy/energy-dispersive X-ray spectrometry (SEM/EDX) analysis was performed on the Au-Ge alloy samples after 24 h of immersion time in artificial sweat and saliva in order to determine the potential changes of the alloying elements on the surface of the alloy samples during the electrochemical measurements. The SEM used was an FEI Sirion 400 NC (FEI Technologies Inc., Hillsboro, OR, USA), equipped with an INCA 350 (Oxford Instruments, Oxfordshire, UK) microchemical spectrometer.

## 3. Results and Discussion

### *3.1. Artificial Sweat*

#### 3.1.1. Open-Circuit Potential Measurement—Artificial Sweat

Figure 1 shows the open-circuit potential (OCP) of the tested Au-Ge alloy as a function of the immersion time in the artificial sweat solution.

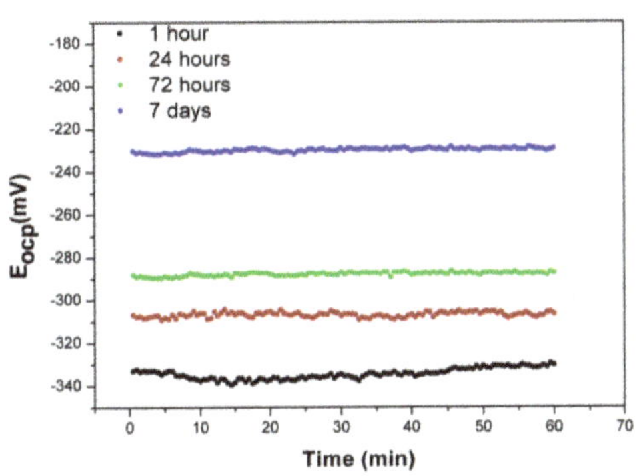


**Figure 1.** Open-circuit potential value as a function of different immersion times.

From Figure 1, it is evident that the open-circuit potential in all the investigated periods was very stable, and, during the given measurement (60 min), did not change significantly. On the other hand, the increase in the immersion time (from one hour to 7 days) caused a significant shift in the open-circuit potential towards the anodic direction (Table 1). The increases in the OCP values in the function of the immersion time and absence of potential drops associated with surface activation suggest that the natural corrosion products formed on the Au-Ge alloy surface are kinetically resistant to chemical dissolution [14,20], and can improve the Au-Ge alloy corrosion protection ability in an artificial sweat solution.

**Table 1.** The corrosion parameters of the Au-Ge alloy at different immersion times in artificial sweat.

| Time | $E_{OCP}$ (mV) | $E_{corr}$ (mV) | $j_{corr}$ ($\mu A/cm^2$) |
|---|---|---|---|
| 1 h | −333 ± 10 | −423 ± 5 | 9.42 ± 1.20 |
| 24 h | −307 ± 8 | −404 ± 7 | 35.2 ± 2.10 |
| 72 h | −289 ± 4 | −383 ± 4 | 17.4 ± 1.50 |
| 7 days | −230 ± 5 | −325 ± 5 | 16.6 ± 1.80 |

3.1.2. Potentiodynamic Measurements—Artificial Sweat

The potentiodynamic polarisation curves of the tested Au-Ge alloy immersed at different times in the artificial sweat solution are shown in Figure 2.

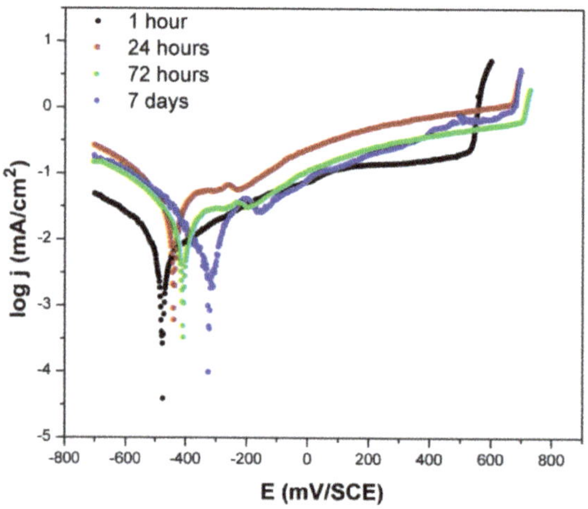


**Figure 2.** Potentiodynamic polarisation curves of the Au-Ge alloy immersed in artificial sweat at different periods.

The corrosion parameters, namely corrosion potential ($E_{corr}$), and corrosion current density ($j_{corr}$), are given in Table 1.

As can be seen from Figure 2 and from the data in Table 1, increasing the immersion time also led to a shift in the corrosion potential in the anodic direction and to a significant increase in the corrosion current density. This kind of behaviour suggests that the layer formed on the alloy's surface in the artificial sweat solution formed a kind of insulating barrier, but it was, nevertheless, permeable to ions' N exchange, which is especially pronounced during the electrode polarisation.

During all measurements, the passive current density can be registered in the anodic part of the polarisation curve up to about 600 mV, when active dissolution started again. The shape of the polarisation curves was the same in all cases; only the length of the Tafel's, mixed (active-diffusion control, slight passivation) and passive region, as well as the value of the passive current density and breakdown potential, changed with the immersion time (Table 2). The shortest passive region was registered after 7 days of immersion, and the highest passive current density was registered after 24 h of immersion (Table 2). This behaviour of the alloy in the artificial sweat solution may indicate that it takes time for the Au-Ge surface to stabilise. On the other hand, the characteristics and protective properties of the Au-Ge alloy surface change slightly over time.

**Table 2.** The passivation parameters of the Au-Ge alloy at different immersion times in artificial sweat.

| Time | Tafel reg. (mV) from $E_{corr}$ to | Mixed reg. (mV) from Tafel reg. to | Passive reg. (mV) from Mix. reg. to | $j_{pass}$ ($\mu A/cm^2$) |
|---|---|---|---|---|
| 1 h | −400 ± 25 | +100 ± 25 | 530 ± 10 | 0.14 ± 0.07 |
| 24 h | −360 ± 15 | +200 ± 10 | 670 ± 5 | 0.80 ± 0.18 |
| 72 h | −320 ± 20 | +200 ± 20 | 700 ± 6 | 0.40 ± 0.20 |
| 7 days | −200 ± 15 | +450 ± 15 | 670 ± 8 | 0.70 ± 0.16 |

#### 3.1.3. EIS Measurements—Artificial Sweat

The impedance spectra of the tested Au-Ge alloy in the artificial sweat solution at different open-circuit potentials are presented as the Bode phase (Figure 3a) and Bode magnitude (Figure 3b).

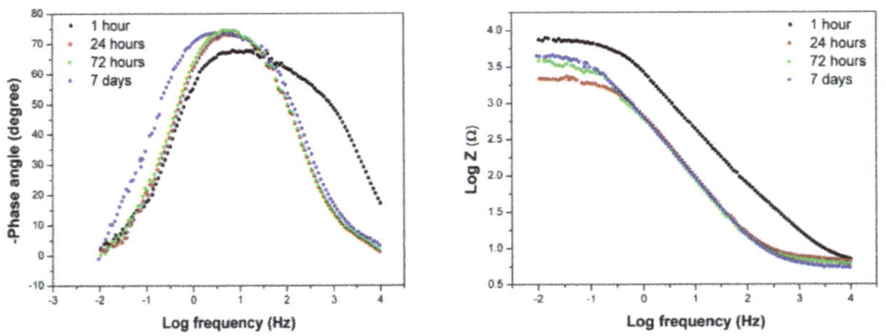


(a) (b)

**Figure 3.** (**a**) The Bode phase angle and (**b**) Bode magnitude of the Au-Ge alloy immersed in artificial sweat at different periods.

The Bode phase plots (Figure 3a) show a deviation in the behaviour of the alloy after 1 h of immersion and a similar behaviour after 24 h. After one hour of immersion in the Bode phase plots, two time constants can be registered, in contrast to the later measurements where only one time constant occurred (after 24 h). One of the time constants was at the low intermediate frequency region (phase angle approaching 0°), where the maximum phase angle (66°) was located at approximately 10 Hz. The second time constant was observed in the high intermediate frequency region with a phase angle which was not approaching 0°, and the maximum phase angle was ~58° at approximately 300 Hz. This time constant, at the intermediate frequency region, may correspond to the presence of the viscous film on the alloy's surface, and could be associated with the formation of a corrosion product on the alloy's surface [12,21,22].

During the immersion, after 24 h, a slight increase in the phase angle was registered, but it did not change further as a function of time. A slight increase in the phase angle at the beginning of the immersion time indicates that it takes time (approximately 24 h) to form and stabilise the actual performance of the Au-Ge alloy surface toward the corrosion process. The phase angle plots revealed only one peak for measurements after 24 h, indicating the involvement of a single time constant that now exists in the system. The maximum phase angles noticed in the range of approximately −70° to −80° in the low frequency range (~5 Hz) are typically characteristic of capacitive behaviour, corresponding to the good corrosion resistance of the materials [23], and are an indication of the presence of a layer on the alloy's surface [14,24,25].

The Bode magnitude (Figure 3b) showed that, in all the measurements in the low frequency's region, the high impedance values (log Z~4–5) for the tested alloys suggested a good corrosion resistance. The lowest impedance value was registered for the immersion time of 24 h. Furthermore, it should also be noted that, in the Bode absolute (Figure 3b) at high frequencies, for an immersion time of one hour, the modulus of impedance was in a decreasing trend, in contrast to the more or less constant modulus of impedance for other immersion times, which is characteristic behaviour based on the literature [26–28]. The non-constant behaviour of the constant modulus at high frequencies certainly confirms that, after one hour in artificial sweat, the film formed on the Au-Ge alloy surface was still rough and porous [23]. In other measurements (except after one hour), the slopes of Z as a function of frequency were approximately −0.8. This means that the protective layer formed on the alloy's surface in the artificial sweat solution formed a kind of insulating barrier, but it was still permeable to ions from the solution [12,29].

Based on the EIS data, it is possible to establish which equivalent circuit is suitable for modelling the electrochemical behaviour of the tested Au-Ge alloy in artificial sweat solutions. Because of the fact that in the impedance spectra (Figure 3a,b), a non-ideal frequency response was evident, a constant phase element (Q) was used in the equivalent circuit, which was then converted to the double layer capacitance using the usual equation [12,23,30]. The obtained impedance spectra were fitted using the $R_s$ ($R_p$ $Q_{dl}$) circuits, except for the 1 h immersion time, where the $R_s$ ($R_p$ $Q_{dl}$) ($R_f$ $C_f$) model was used. Table 3 gives the values of electrochemical parameters as follows: $R_s$ (solution resistance), $R_p$ (polarisation resistance), $C_{dl}$ (double layer capacitance), $R_f$ (film resistance), $C_f$ (film capacitance), and n (constant phase exponent) obtained from the fitting of the EIS data.

**Table 3.** EIS parameters of the Au-Ge alloy at different immersion times in artificial sweat.

| Time | $R_s$ (Ohm/cm$^2$) | $R_p$ (kOhm/cm$^2$) | $C_{dl}$ (mF/cm$^2$) | n | $R_f$ (kOhm/cm$^2$) | $C_f$ (μF/cm$^2$) | $\chi^2 \cdot 10^{-3}$ |
|---|---|---|---|---|---|---|---|
| 1 h | 15 ± 1.0 | 5.80 ± 0.02 | 1.05 ± 0.10 | 0.85 ± 0.005 | 2.0 ± 0.2 | 2.0 ± 0.02 | 2.20 |
| 24 h | 5 ± 0.5 | 2.15 ± 0.03 | 3.67 ± 0.11 | 0.87 ± 0.001 | - | - | 1.15 |
| 72 h | 5 ± 0.5 | 4.53 ± 0.10 | 4.55 ± 0.21 | 0.84 ± 0.002 | - | - | 2.82 |
| 7 days | 5 ± 0.5 | 4.85 ± 0.08 | 4.45 ± 0.18 | 0.85 ± 0.002 | - | - | 1.87 |

As already stated, the immersion time of 1 h differs in the behaviour when compared to other times, and was described by a model with two time constants (Table 3). The second time constant relates to the presence of a film or corrosion products on the alloy's surface. As can be seen from the results (Table 3) in periods from 24 h to 7 days, with the increase in the immersion time, there was a slight increase in the polarisation resistance ($R_p$) and the double layer capacitance ($C_{dl}$). The obtained results, as in the case of the polarisation measurements, indicate that it takes time to stabilise the Au-Ge alloy's surface and form permanent properties. Between the immersion times of 72 h and 7 days, there was no noticeable difference in the alloy behaviour.

### 3.2. *Artificial Saliva*

#### 3.2.1. Open-Circuit Potential Measurement—Artificial Saliva

As in the case of the investigation with the artificial sweat solution, the open-circuit potential was very stable during the tested period in the artificial saliva, and did not change significantly during the given measurement (60 min). Increasing the immersion time (from one hour to 7 days) caused a significant shift in the open-circuit potential, unlike the sweat solution, towards the cathodic direction (Table 4). This kind of behaviour suggests that the layer formed on the alloy's surface blocked the active cathodic part, and therefore the reduction reaction was shifted to a more negative potential range.

**Table 4.** The corrosion parameters of the Au-Ge alloy at different immersion times in artificial saliva.

| | pH = 4.81 | | | pH = 2.51 | | |
|---|---|---|---|---|---|---|
| Time | $E_{OCP}$ (mV) | $E_{corr}$ (mV) | $j_{corr}$ (μA/cm$^2$) | $E_{OCP}$ (mV) | $E_{corr}$ (mV) | $j_{corr}$ (μA/cm$^2$) |
| 1 h | −24 ± 5 | 96 ± 10 | 0.30 ± 0.15 | 20 ± 8 | 112 ± 10 | 0.26 ± 0.10 |
| 24 h | −50 ± 3 | 8 ± 5 | 0.15 ± 0.09 | −362 ± 10 | −361 ± 10 | 12.94 ± 1.85 |
| 72 h | −268 ± 7 | −247 ± 9 | 1.56 ± 0.50 | −367 ± 7 | −372 ± 7 | 14.09 ± 1.26 |
| 7 days | −339 ± 4 | −320 ± 4 | 1.95 ± 0.80 | −376 ± 9 | −388 ± 9 | 9.37 ± 0.98 |

### 3.2.2. Potentiodynamic Measurements—Artificial Saliva

The potentiodynamic polarisation curves of the tested Au-Ge alloy immersed in the artificial saliva solution at pH 4.81 and 2.51 for different times are shown in Figures 4 and 5, respectively. The corrosion parameters, namely the corrosion potential ($Ec_{orr}$) and corrosion current density ($j_{corr}$), are given in Table 4.

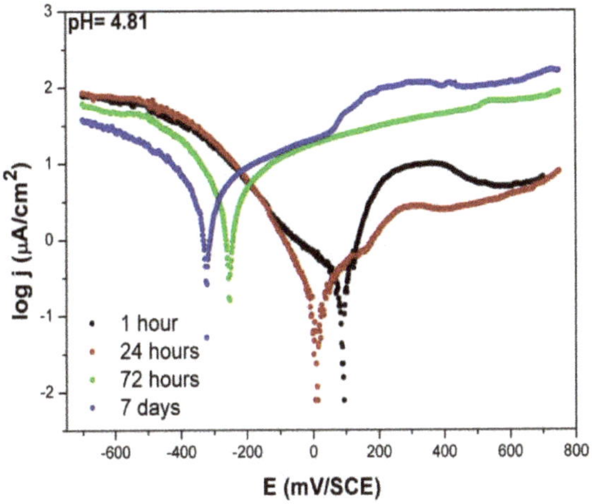


**Figure 4.** Potentiodynamic polarisation curves of the Au-Ge alloy immersed in artificial saliva (pH = 4.81) at different periods.

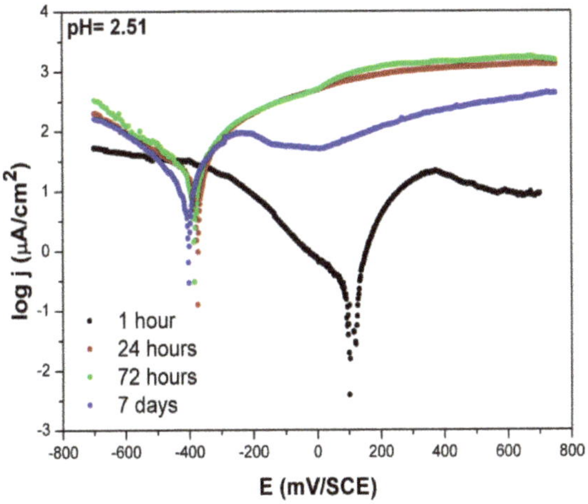


**Figure 5.** Potentiodynamic polarisation curves of the Au-Ge alloy immersed in artificial saliva (pH = 2.51) at different periods.

As can be seen from Figures 4 and 5 and from the data shown in Table 4, increasing the immersion time led to a shift in the corrosion potential in the cathodic direction and an increase in the corrosion current density in both saliva solutions (pH = 4.81 and pH = 2.51). The difference was that the balance was established much faster in the acidic solution. Namely, after 24 h, no major changes in the corrosion potential were observed, while, in the original saliva solution (pH = 4.81), it took longer. The corrosion potential in both solutions after 7 days was not much different, but, in the acidic saliva solution, the corrosion current density was significantly higher than in the original artificial saliva solution (Table 4).

As in the case of the artificial sweat solution, the expressed passive current density was obtained in the anodic part of the polarisation curve. In all the measurements, after the Tafel's region, a passive region appeared up to the range of measurements we performed (Table 5). In both saliva solutions, an increase in the passive current density was registered when increasing the immersion time. In the acidic saliva solution, on the other hand, the passive area started at a significantly more negative potential, but with a much higher passive current density (Table 5).

**Table 5.** The passivation parameters of the Au-Ge alloy at different immersion times in artificial saliva.

| | pH = 4.81 | | pH = 2.51 | |
|---|---|---|---|---|
| **Time** | **Tafel reg. (mV) from $E_{corr}$ to** | $j_{pass}$ **(μA/cm²)** | **Tafel reg. (mV) from $E_{corr}$ to** | $j_{pass}$ **(μA/cm²)** |
| 1 h | 230 ± 12 | 5.0 ± 2.0 | 370 ± 20 | 9.0 ± 1.8 |
| 24 h | 280 ± 25 | 3.2 ± 1.8 | −270 ± 18 | 1100 ± 50 |
| 72 h | −100 ± 14 | 39 ± 7.0 | −270 ± 16 | 1400 ± 75 |
| 7 days | −220 ± 8 | 110 ± 2.0 | −240 ± 10 | 250 ± 10 |

3.2.3. EIS Measurements—Artificial Saliva

The impedance spectra of the Au-Ge alloy measured at the open-circuit potential, after immersion at different times in the artificial saliva solution at pH 4.81, are shown as the Bode phase and Bode magnitude in Figures 6a and 6b, respectively, and at pH 2.51 in Figures 7a and 7b, respectively. The impedance parameters are given in Table 6.

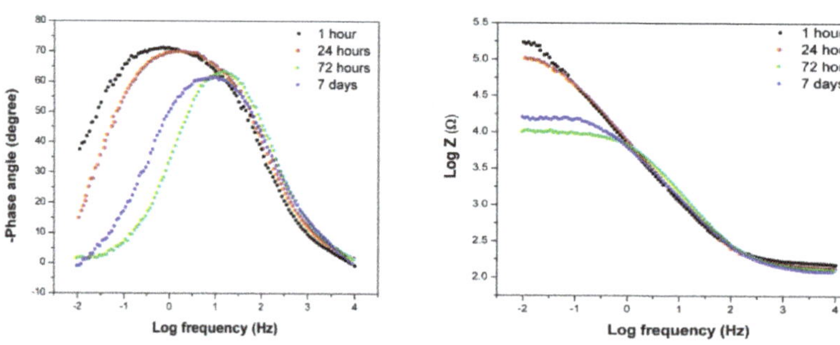


(**a**) (**b**)

**Figure 6.** (**a**) The Bode phase angle and (**b**) Bode magnitude of the Au-Ge alloy immersed in artificial saliva (pH = 4.81) at different periods.

From the data shown (Figures 6a and 7a), in both the tested saliva solutions, independent of the immersion time, only one time constant appeared, and that was at the low intermediate frequency region. In the original saliva solution (pH = 4.81), the highest value of the phase angle, when compared to the acidic solution, was registered (Table 3). In the original saliva solution, a slight shift in the phase maximum as a function of immersion time can be registered towards higher frequencies, but the maximum value remained in the intermediate frequency region (Figure 6a). As mentioned earlier, this frequency region (the intermediate frequency region) describes the capacitive behaviour corresponding to the corrosion resistance of materials [23], and was an indication of the presence of a layer on the alloy's surface [14,24].

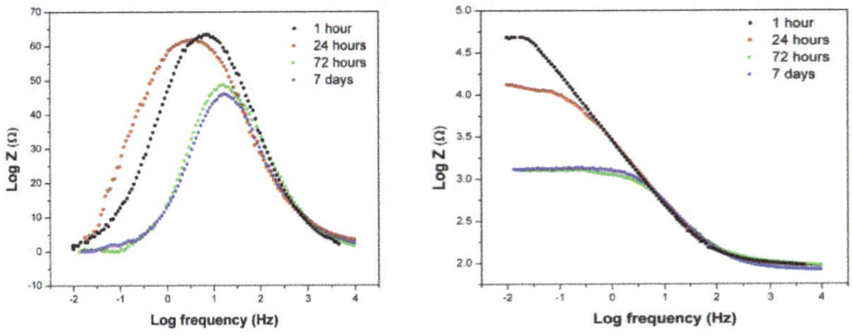


(**a**) (**b**)

**Figure 7.** (**a**) The Bode phase angle and (**b**) Bode magnitude of the Au-Ge alloy immersed in artificial saliva (pH = 2.51) at different periods.

**Table 6.** The impedance and phase value of the Au-Ge alloy at different immersion times in artificial saliva.

| | pH = 4.81 | | pH = 2.51 | |
|---|---|---|---|---|
| **Time** | **Log Z** | **−phase** | **Log Z** | **−phase** |
| 1 h | 5.2 | 72 | 4.7 | 63 |
| 24 h | 5.2 | 70 | 4.0 | 62 |
| 72 h | 4.2 | 64 | 3.1 | 48 |
| 7 days | 4.0 | 62 | 3.0 | 46 |

The lower phase angle values indicated that the corrosion resistance of the tested alloy decreased with the increasing immersion time, as well as a decrease in the pH values.

In the high frequency region, all the Bode magnitudes (Figures 6b and 7b) showed a plateau when the phase angle approached 0°, which is an indication of electrolyte resistance.

In the intermediate frequency region, the linear slope in log Z as a function of log (f) was approximately −0.8 for the original saliva solution (pH = 4.81), with the phase angle decreasing as a function of time from 72° to 62°. In the case of the acidic saliva solution (pH = 2.51), the linear slope was significantly lower, about −0.5, with an additional lower phase angle that decreased as a function of time from 63° to 46°. A slope of −0.5 and phase angle of about 45° is characteristic of the system via diffusion control [12] (acidic saliva solution after 72 h, Table 6). This kind of behaviour indicates that the corrosion resistance of the tested alloy was not perfect in the artificial saliva, and that it decreased further during the immersion time. In the acidic saliva solution (pH = 2.51), the decrease was more pronounced.

In the low frequency region, in the case of all measurements, another plateau was registered, with the phase angle approaching 0° (except for the immersion times of 1 h and 24 h at pH = 4.81), which describes the charge transfer processes. The deviation of the behaviour (immersion times of 1 and 24 h at pH = 4.81) may indicate the existence of processes in the pores of the surface film [12]. Since this deviation disappeared with time, it may be a sign that the time of 24 h was insufficient for the Au-Ge alloy surface to stabilise and form permanent properties.

As in the case of the artificial sweat solution, the obtained impedance spectra in the saliva solution were fitted using the $R_s$ ($R_pQ_{dl}$) circuits. The electrochemical parameters $R_s$, $R_p$, $C_{dl}$, and n, obtained from the fitting of the EIS data, are given in Table 7.

**Table 7.** EIS parameters of the Au-Ge alloy at different immersion times in artificial saliva.

| | pH = 4.81 | | | | | pH = 2.51 | | | | |
|---|---|---|---|---|---|---|---|---|---|---|
| Time | $R_s$ ($Ohm/cm^2$) | $R_p$ ($kOhm/cm^2$) | $C_{dl}$ ($mF/cm^2$) | n | $\chi^2 \cdot 10^{-3}$ | $R_s$ ($Ohm/cm^2$) | $R_p$ ($kOhm/cm^2$) | $C_{dl}$ ($mF/cm^2$) | n | $\chi^2 \cdot 10^3$ |
| 1 h | 5 ± 1 | 280 ± 0.2 | 0.43 ± 0.01 | 0.84 ± 0.002 | 5.2 | 5 ± 2 | 270 ± 0.4 | 1.10 ± 0.02 | 0.83 ± 0.001 | 1.1 |
| 24 h | 5 ± 1 | 120 ± 1.0 | 0.37 ± 0.02 | 0.84 ± 0.001 | 2.5 | 80 ± 5 | 2.50 ± 0.1 | 1.24 ± 0.017 | 0.81 ± 0.001 | 4.3 |
| 72 h | 15 ± 0.2 | 10.25 ± 0.5 | 0.24 ± 0.01 | 0.80 ± 0.003 | 1.8 | 80 ± 4 | 1.30 ± 0.08 | 2.37 ± 0.032 | 0.86 ± 0.001 | 5.9 |
| 7 days | 15 ± 0.2 | 16.00 ± 0.4 | 0.11 ± 0.008 | 0.80 ± 0.001 | 2.2 | 80 ± 5 | 1.20 ± 0.15 | 2.51 ± 0.022 | 0.81 ± 0.002 | 2.1 |

For both saliva solutions, an increase in the immersion time led to a considerable decrease in the polarisation resistance ($R_p$) of the tested Au-Ge alloy, which was much more pronounced in the case of the acidic saliva solution (Table 7). The difference appeared in the tendency to change the double layer capacitance ($C_{dl}$). Namely, in the case of the original saliva (pH = 4.81), with increasing time, the value of $C_{dl}$ decreased slightly. On the other hand, increasing the time in the case of an acidic saliva solution caused an increase in the value of $C_{dl}$, similar to the case of the sweat solution. This behaviour indicates the possibility of a higher ion exchange in the double layer, due to which the corrosion resistance of the alloy is reduced significantly in the acidic saliva solution when compared to the original saliva solution.

### 3.3. SEM/EDX Analysis

The SEM/EDX analysis was performed on samples after 24 h of immersion in artificial sweat and saliva to monitor the changes of the alloying elements on the surfaces of the alloy samples during electrochemical measurements. The normalised values of the alloy surface's constituent elements are shown in Figure 8.

Artificial sweat | Artificial saliva

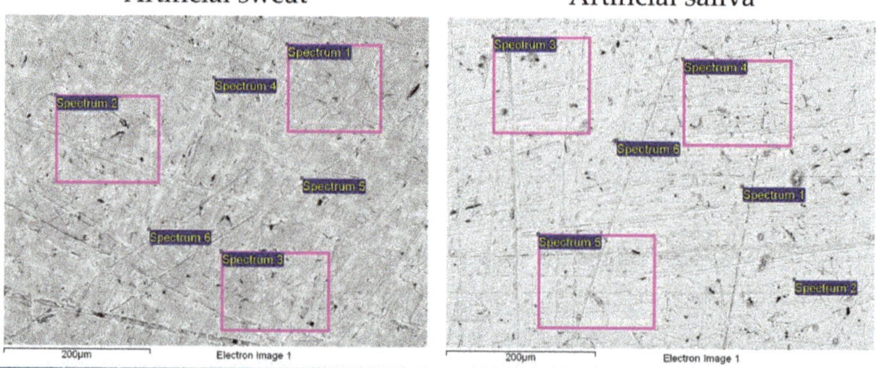


Artificial sweat:

| Spectrum | O | Cu | Zn | Ge | Au | Total |
|---|---|---|---|---|---|---|
| Spectrum 1 | 1.95 | 25.90 | 3.80 | 1.69 | 66.66 | 100.00 |
| Spectrum 2 | 2.52 | 23.42 | 3.17 | 1.33 | 69.57 | 100.00 |
| Spectrum 3 | 1.82 | 24.80 | 3.29 | 0.63 | 69.46 | 100.00 |
| Spectrum 4 | 1.25 | 23.34 | 3.02 | 1.83 | 70.56 | 100.00 |
| Spectrum 5 | 2.42 | 20.30 | 2.81 | 1.42 | 73.05 | 100.00 |
| Spectrum 6 | 0.50 | 25.48 | 3.32 | 0.99 | 69.70 | 100.00 |
| Max. | 2.52 | 25.90 | 3.80 | 1.83 | 73.05 | |
| Min. | 0.50 | 20.30 | 2.81 | 0.63 | 66.66 | |
| Std. deviation | 0.76 | 2.04 | 0.33 | 0.44 | 2.06 | |
| Mean | 1.74 | 23.87 | 3.24 | 1.32 | 69.83 | 100.00 |

Artificial saliva:

| Spectrum | O | Cu | Zn | Ge | Au | Total |
|---|---|---|---|---|---|---|
| Spectrum 1 | 2.03 | 29.80 | 4.46 | 1.70 | 62.00 | 100.00 |
| Spectrum 2 | 0.70 | 31.29 | 4.58 | 1.15 | 62.27 | 100.00 |
| Spectrum 3 | 3.52 | 28.78 | 4.24 | 1.62 | 61.83 | 100.00 |
| Spectrum 4 | 3.64 | 29.35 | 4.07 | 1.59 | 61.34 | 100.00 |
| Spectrum 5 | 3.12 | 30.36 | 4.25 | 1.66 | 60.60 | 100.00 |
| Spectrum 6 | 2.04 | 30.31 | 4.39 | 2.19 | 61.06 | 100.00 |
| Max. | 3.64 | 31.29 | 4.58 | 2.19 | 62.27 | |
| Min. | 0.70 | 28.78 | 4.07 | 1.15 | 60.60 | |
| Std. deviation | 1.13 | 0.88 | 0.18 | 0.33 | 0.63 | |
| Mean | 2.51 | 29.98 | 4.33 | 1.65 | 61.52 | 100.00 |

**Figure 8.** SEM/EDX analysis of the Au-Ge alloy samples after 24 h of immersion in the artificial sweat and artificial saliva solution during electrochemical testing.

Oxygen was detected with the EDX analysis, showing the oxidation of the sample surfaces during electrochemical testing. The sample exposed to artificial sweat had a somewhat lower Cu content and higher Au content as compared to the sample exposed to the artificial saliva. In correlation with the EIS measurements, the difference in composition on the alloy sample surface may correspond to the formation of the corrosion-resistant layer. The less noble elements of Cu and Zn may have been dissociated into the acidic surrounding medium more easily, while the Au remained on the sample surface. The corrosion products formed on the alloy surface had a higher Au content, which was inherently more corrosion-resistant, making the surface layer less susceptible to continued chemical dissolution. In the artificial saliva solution, the metallic element dissociation took longer to form the final film on the alloy's surface, and the EDX analysis showed only a slightly altered Cu and Au content from the initial chemical alloy composition.

The SEM images show several scratches and dark spots. The scratches are the result of surface preparation before exposure to the artificial media, while the dark spots are mostly contamination and impurities from handling the samples with artificial saliva and sweat. Some dark spots are burrs and edges of the scratches, holes, or imperfections on the sample surface. Examining the effects of exposure to the artificial media has shown that very little corrosion damage was obtained on the sample surfaces.

## 4. Conclusions

The electrochemical behaviour of the Au-Ge alloy at different immersion times in the artificial sweat and artificial saliva solutions (different pH values) was studied using potentiodynamic polarisation measurements, electrochemical impedance spectroscopy, and SEM/EDX analysis.

Artificial sweat solution:

- The obtained results showed that the polarisation curves of the tested alloy were similar in all measurements, suggesting that the main corrosion mechanism was the same, regardless of the immersion times. The difference in behaviour occurred only in the length of the Tafel's, mixed, and passive region, as well as in the value of the passive current density.
- As a function of immersion time, a shift in $E_{OCP}$ and $E_{corr}$ towards the anodic direction was also recorded, as well as an increase in the corrosion current density. This kind of behaviour suggests that the layer formed on the alloy's surface in the artificial sweat solution formed a kind of insulating barrier, but it was, nevertheless, permeable to ions' exchange.
- The EIS measurements confirmed that, after one hour in the artificial sweat solution, the Au-Ge alloy surface did not stabilise. Because of that, the immersion time of 1 h differed in the behaviour when compared to the later period, and was described by a model with two time constants in contrast to the later measurements, where only one time constant occurred.
- The slight increase in the polarisation resistance and the double layer capacitance over time also suggests that the layer formed on the alloy's surface in the artificial sweat solution behaved as a kind of insulating barrier, but it was still permeable to ions from the solution.

Artificial saliva solution:

- The potentiodynamic polarisation measurement showed that, in both saliva solutions during the immersion time, a shift in the $E_{OCP}$ and $E_{corr}$ towards the cathodic direction was recorded, as well as an increase in the corrosion current density.
- After 7 days, the corrosion potential in both solutions was not much different, but, in the acidic saliva solution, the corrosion current density was significantly higher (higher corrosion rate) than in the original artificial saliva solution.
- Only one time constant appeared in the EIS measurements of both tested saliva solutions, independent of the immersion time.

- The results obtained at different immersion times showed that the corrosion resistance of the tested alloy decreased as a function of time.
- The results also showed that, in the acidic saliva solution (pH = 2.51), the corrosion rate of the studied alloy was higher, which was accompanied by a decrease in the impedance (Z), phase angle, and polarisation resistance, and an increase in the double layer capacitance over time when compared to the original saliva solution (pH = 4.81).
- The SEM/EDX analysis results are in good agreement with the results obtained through the electrochemical measurements in the artificial sweat and the artificial saliva solutions.

**Author Contributions:** Conceptualisation, G.V. and R.R.; methodology, G.V., P.M. and R.R.; software, G.V. and P.M.; validation, G.V., V.L. and R.R.; formal analysis, G.V. and P.M.; investigation, G.V., R.R. and P.M.; resources, V.L. and R.R.; writing—original draft preparation, G.V. and P.M.; writing—review and editing, R.R.; supervision, R.R.; project administration, P.M.; funding acquisition, V.L. and R.R. All authors have read and agreed to the published version of the manuscript.

**Funding:** The authors gratefully acknowledge the financial support of the Ministry of Science, Technological Development and Innovation of the Republic of Serbia (Grants Nos. 451-03-66/2024-03/200125 and 451-03-65/2024-03/200125) and The Bilateral Projects Serbia—Slovenia (337-00-110/2023-05/26).

**Data Availability Statement:** The original contributions presented in the study are included in the article, further inquiries can be directed to the corresponding author.

**Conflicts of Interest:** Authors Peter Majerič and Rebeka Rudolf were employed by the company Zlatarna Celje d.o.o. The remaining authors declare that the research was conducted in the absence of any commercial or financial relationships that could be construed as a potential conflict of interest.

## References


1. Nikpour, S.; Henderson, J.D.; Matin, S.; Nie, H.-Y.; Hedberg, J.; Dehnavi, V.; Hosein, Y.K.; Holdsworth, D.W.; Biesinger, M.; Hedberg, Y.S. Effect of Passivation and Surface Treatment of a Laser Powder Bed Fusion Biomedical Titanium Alloy on Corrosion Resistance and Protein Adsorption. *Electrochim. Acta* **2024**, *475*, 143650. [CrossRef]
2. Atapour, M.; Sanaei, S.; Wei, Z.; Sheikholeslam, M.; Henderson, J.D.; Eduok, U.; Hosein, Y.K.; Holdsworth, D.W.; Hedberg, Y.S.; Ghorbani, H.R. In Vitro Corrosion and Biocompatibility Behavior of CoCrMo Alloy Manufactured by Laser Powder Bed Fusion Parallel and Perpendicular to the Build Direction. *Electrochim. Acta* **2023**, *445*, 142059. [CrossRef]
3. Caporali, S.; Bardi, U. Corrosion Mechanism in Artificial Sweat Solution of In-Bearing White Bronze Alloy. *Corrosion* **2012**, *68*, 025001-1–025001-8. [CrossRef]
4. Markel, K.; Silverberg, N.; Pelletier, J.L.; Watsky, K.L.; Jacob, S.E. Art of Prevention: A Piercing Article about Nickel. *Int. J. Women's Dermatol.* **2020**, *6*, 203–205. [CrossRef] [PubMed]
5. Rudolf, R.; Majerič, P.; Lazić, V.; Grgur, B. Development of a New AuCuZnGe Alloy and Determination of Its Corrosion Properties. *Metals* **2022**, *12*, 1284. [CrossRef]
6. Cho, J.M.; Chae, J.; Jeong, S.R.; Moon, M.J.; Shin, D.Y.; Lee, J.H. Immune Activation of Bio-Germanium in a Randomized, Double-Blind, Placebo-Controlled Clinical Trial with 130 Human Subjects: Therapeutic Opportunities from New Insights. *PLoS ONE* **2020**, *15*, e0240358. [CrossRef] [PubMed]
7. Ji, Y.; Yang, S.; Sun, J.; Ning, C. Realizing Both Antibacterial Activity and Cytocompatibility in Silicocarnotite Bioceramic via Germanium Incorporation. *J. Funct. Biomater.* **2023**, *14*, 154. [CrossRef] [PubMed]
8. Zemek, J.; Jiricek, P.; Houdkova, J.; Ledinsky, M.; Jelinek, M.; Kocourek, T. On the Origin of Reduced Cytotoxicity of Germanium-Doped Diamond-Like Carbon: Role of Top Surface Composition and Bonding. *Nanomaterials* **2021**, *11*, 567. [CrossRef]
9. Wang, Y.; Teng, G.; Zhou, H.; Dong, C. Germanium Reduces Inflammatory Damage in Mammary Glands During Lipopolysaccharide-Induced Mastitis in Mice. *Biol. Trace Elem. Res.* **2020**, *198*, 617–626. [CrossRef]
10. Elango, J.; Bushin, R.; Lijnev, A.; De Aza, P.N.; Martínez, C.P.-A.; Marín, J.M.G.; Hernandez, A.B.; Olmo, L.R.M.; Val, J.E.M.S. De The Effect of Germanium-Loaded Hydroxyapatite Biomaterials on Bone Marrow Mesenchymal Stem Cells Growth. *Cells* **2022**, *11*, 2993. [CrossRef]
11. Kurt, M.Ş.; Arslan, M.E.; Yazici, A.; Mudu, İ.; Arslan, E. Tribological, Biocompatibility, and Antibiofilm Properties of Tungsten–Germanium Coating Using Magnetron Sputtering. *J. Mater. Sci. Mater. Med.* **2021**, *32*, 6. [CrossRef] [PubMed]
12. Porcayo-Calderon, J.; Rodríguez-Díaz, R.A.; Porcayo-Palafox, E.; Martinez-Gomez, L. Corrosion Performance of Cu-Based Coins in Artificial Sweat. *J. Chem.* **2016**, *2016*, 9542942. [CrossRef]
13. Chen, Y.-L.; Kuan, W.-H.; Liu, C.-L. Comparative Study of the Composition of Sweat from Eccrine and Apocrine Sweat Glands during Exercise and in Heat. *Int. J. Environ. Res. Public Health* **2020**, *17*, 3377. [CrossRef] [PubMed]

14. Mareci, D.; Chelariu, R.; Dan, I.; Gordin, D.-M.; Gloriant, T. Corrosion Behaviour of β-Ti20Mo Alloy in Artificial Saliva. *J. Mater. Sci. Mater. Med.* **2010**, *21*, 2907–2913. [CrossRef] [PubMed]
15. Lima, A.R.; Pinto, A.M.P.; Toptan, F.; Alves, A.C. Impact of Simulated Inflammation and Food Breakdown on the Synergistic Interaction between Corrosion and Wear on Titanium. *Corros. Sci.* **2024**, *228*, 111839. [CrossRef]
16. Bodunrin, M.O.; Chown, L.H.; Van Der Merwe, J.W.; Alaneme, K.K.; Oganbule, C.; Klenam, D.E.P.; Mphasha, N.P. Corrosion Behavior of Titanium Alloys in Acidic and Saline Media: Role of Alloy Design, Passivation Integrity, and Electrolyte Modification. *Corros. Rev.* **2020**, *38*, 25–47. [CrossRef]
17. Begum, A.A.S.; Vahith, R.M.A.; Mohamed, M.K.V.; Kotra, V.; Shaik, B.; Al-Kahtani, A. Corrosion Mitigation on Orthodontic Wire Made of SS 18/8 Alloy Using Esomeprazole Tablet (Esiloc-40 Mg) in Artificial Saliva. *J. Saudi Chem. Soc.* **2023**, *27*, 101681. [CrossRef]
18. *ISO 3160-2:2015*; Watch-Cases and Accessories—Gold Alloy Coverings. International Organization for Standardization: Geneva, Switzerland, 2015.
19. Rathish, R.J.; Rajendran, S.; Christy, J.L.; Devi, B.S.; Johnmary, S.; Manivannan, M.; Rajam, K.; Rengan, P. Corrosion Behaviour of Metals in Artificial Sweat. *Open Corros. J.* **2010**, *3*, 38–44. [CrossRef]
20. Jimenez-Marcos, C.; Mirza-Rosca, J.C.; Baltatu, M.S.; Vizureanu, P. Effect of Si Contents on the Properties of Ti15Mo7ZrxSi Alloys. *Materials* **2023**, *16*, 4906. [CrossRef]
21. Porcayo-Calderon, J.; Rodriguez-Diaz, R.A.; Porcayo-Palafox, E.; Colin, J.; Molina-Ocampo, A.; Martinez-Gomez, L. Effect of Cu Addition on the Electrochemical Corrosion Performance of $Ni_3Al$ in 1.0 M $H_2SO_4$. Adv. *Mater. Sci. Eng.* **2015**, *2015*, 209286. [CrossRef]
22. Canto, J.; Rodríguez-Díaz, R.A.; Martinez-de-la-Escalera, L.M.; Neri, A.; Porcayo-Calderon, J. Corrosion Inhibition in $CO_2$-Saturated Brine by Nd3+ Ions. *Molecules* **2023**, *28*, 6593. [CrossRef] [PubMed]
23. Dutta Chowdhury, N.; Ghosh, K.S. Electrochemical Behaviour of Dental Amalgam in Natural, Artificial Saliva and in 0.90 Wt.% NaCl Solution. *Corros. Sci.* **2018**, *133*, 217–230. [CrossRef]
24. Falodun, O.E.; Oke, S.R.; Solomon, M.M.; Bayode, A. Microstructural Evolution and Electrochemical Corrosion Characteristics of Ti–Ni Matrix Composite in NaCl and HCl Solutions. *Ceram. Int.* **2024**, *50*, 15124–15133. [CrossRef]
25. Torbati-Sarraf, H.; Ding, L.; Khakpour, I.; Daviran, G.; Poursaee, A. Unraveling the Corrosion of the Ti–6Al–4V Orthopedic Alloy in Phosphate-Buffered Saline (PBS) Solution: Influence of Frequency and Potential. *Corros. Mater. Degrad.* **2024**, *5*, 276–288. [CrossRef]
26. Fekry, A.M.; El-Sherif, R.M. Electrochemical Corrosion Behavior of Magnesium and Titanium Alloys in Simulated Body Fluid. *Electrochim. Acta* **2009**, *54*, 7280–7285. [CrossRef]
27. Arrieta-González, C.D.; Perez-Arizmendi, F.J.; Dorta-Leon, M.A.; Porcayo-Calderón, J. Effect of Temperature on the Corrosion Resistance of Ni5Al Coating Deposited by Electric Arc in 3.5% NaCl Solution. *Coatings* **2023**, *13*, 1349. [CrossRef]
28. Alves, V.A.; Reis, R.Q.; Santos, I.C.B.; Souza, D.G.; de Gonçalves, F.T.; Pereira-da-Silva, M.A.; Rossi, A.; da Silva, L.A. In Situ Impedance Spectroscopy Study of the Electrochemical Corrosion of Ti and Ti–6Al–4V in Simulated Body Fluid at 25 °C and 37 °C. *Corros. Sci.* **2009**, *51*, 2473–2482. [CrossRef]
29. Matemadombo, F.; Nyokong, T. Characterization of Self-Assembled Monolayers of Iron and Cobalt Octaalkylthiosubstituted Phthalocyanines and Their Use in Nitrite Electrocatalytic Oxidation. *Electrochim. Acta* **2007**, *52*, 6856–6864. [CrossRef]
30. Shao, P.; Xiao, H.; Liu, K.; Chen, X.; Hou, M.; Zhang, Q.; Qian, C.; Huang, S. Influence of Annealing Temperature on the Mechanical Properties and Corrosion Behavior of Ti-5.5Al-2.0Zr-1.5Sn-0.5Mo-1.5Nb Alloy. *Arab. J. Chem.* **2024**, *17*, 105790. [CrossRef]

*metals*

MDPI

*Article*

# Mineralogical Characterisation of Copper Slag and Phase Transformation after Carbocatalytic Reduction for Hydrometallurgical Extraction of Copper and Cobalt

**Tina Chanda Phiri, Pritam Singh and Aleksandar N. Nikoloski ***

Centre for Water Energy and Waste, Harry Butler Institute, School of Engineering and Energy, Murdoch University, 90 South Street, Murdoch, WA 6150, Australia; t.phiri@murdoch.edu.au (T.C.P.); p.singh@murdoch.edu.au (P.S.)

* Correspondence: a.nikoloski@murdoch.edu.au

**Abstract:** Copper smelting slag is a significant potential resource for cobalt and copper. The recovery of copper and cobalt from copper slag could significantly augment the supply of these metals, which are essential to facilitating the transition to green energy while simultaneously addressing environmental concerns regarding slag disposal. However, the complex mineral composition of copper slag poses an enormous challenge. This study investigated the mineralogical and chemical characteristics of copper slag, which are vital for devising the most effective processing techniques. XRD and FESEM-EDS were employed to examine the morphologies of copper slag before and after the reduction process. The effects of borax and charcoal (carbocatalytic) reduction on phase transformation were evaluated. The XRD analysis revealed that the primary phases in the copper slag were $Fe_2SiO_4$ and $Fe_3O_4$. The FESEM-EDS analysis verified the presence of these phases and yielded supplementary details regarding metal embedment in the $Fe_2SiO_4$, $Fe_3O_4$, and Cu phases. The carbocatalytic reduction process expedited the transformation of copper slag microstructures from crystalline dendritic to amorphous and metallic phases. Finally, leaching experiments demonstrated the potential benefits of carbocatalytic reduction by yielding high extractions of Cu, Co, and Fe.

**Keywords:** copper slag; carbocatalytic reduction; mineralogical characterisation; phase transformation

## 1. Introduction

Confronted with increasing industrialisation and civilisation, humanity is constantly seeking innovative technologies for a sustainable future. As a result, the extraction of mineral resources has progressively escalated to keep pace with the rapid development of society, and a few of these resources are nearing exhaustion. As high-grade deposits deplete, larger amounts of ore must be mined to provide an adequate supply of raw mAterials, resulting in increased generation of metallurgical waste. Copper smelting slag is a significant metallurgical waste, with a global annual production of 37.7 million tonnes [1]. The disposal of these substantial volumes of slag is concerning due to the potential release of hazardous substances into the environment. Copper slag waste contains hazardous metals, which mAy be a source of pollution affecting surface water, soil, and groundwater [2–4]. This could have an adverse impact on human health and the loss of biodiversity.

Conversely, smelting slag contains a substantial quantity of valuable metals, mAking it a highly appealing alternative mineral resource to meet the growing demand for critical metals in social and economic development [2,3]. Therefore, it is increasingly necessary to reduce waste disposal by extracting valuable metals from the slag waste, thereby contributing to a sustainable future. However, unlike the primary ore, the complex mineralogy of copper slag waste, which primarily consists of mAgnetite and fayalite, mAkes it extremely difficult to extract metals from it [5–7]. Fayalite ($2FeO{\cdot}SiO_2$) is a stoichiometric compound that melts at 1208 °C [8]. The significant presence of fayalite in copper slag poses huge processing challenges due to its high melting point, necessitating energy-intensive processing [9–11]. As a result, a significant amount of slag waste is still stored, which leads to the buildup of slag waste outside of smelters and wastes vital resources while also polluting the environment. Morphology, mineralogical characteristics, and metal distribution are critical factors to understanding the resource utilisation potential of copper smelting slag.

The mineralogical composition of copper slag and the associated distribution of heavy metals were studied by Zhang et al. [6]. Sequential extraction demonstrated that fayalite was more difficult to extract than mAgnetite and hedenbergite due to its significantly stronger binding ability with heavy metals [6]. Guo et al. [12] discovered that during the pyrometallurgical smelting of copper concentrates, heavy metals like Cu, Pb, and As interact with $Fe_2SiO_4$ to form a compact encapsulation state. Several studies discovered that copper slag contains various valuable metals dispersed across different phases [2,13,14]. Our previous study has demonstrated that copper slag contains substantial quantities of valuable metals dispersed in other phases, including 2.59 wt.% Al, 1.83 wt.% Zn, 1.67 wt.% Ag, 1.19 wt.% Cu, and 0.48 wt.% Co [3]. The mineralogical composition of copper slag is complicated by the presence of these valuable metals in the mAjority of stable oxides and silicates. [6] observed that the tightly bound state of hazardous metals in various phases is indistinct, mAking the identification of the primary bound phases of hazardous metals more challenging. Nevertheless, only a limited number of studies have comprehensively analysed the microstructure and predominant occurrence of metal phases in copper slag. Additionally, to date, there has been no research conducted on the phase transformation of copper slag using charcoal in the presence of borax (carbocatalytic reduction). The objective of this study was to examine the morphology of copper slag and phase transformation produced by carbocatalytic reduction (CCR), which is fundamental for optimising the efficiency of disintegration and metal extraction. The research involved the use of microscopic and spectroscopic techniques, as well as laboratory chemistry experiments. The morphology of and phases present in the copper slag were determined using X-ray diffraction (XRD). The mineral phases of Co, Cu, Ti, Ca, Mg, Al, Fe, and Si were identified, and their distribution was analysed using a field emission scanning electron microscope equipped with a spectrometer with X-ray energy dispersion (FESEM-EDS). Additionally, the potential effect of carbocatalytic reduction for enhanced metal extraction was assessed through leaching experiments on the product of the copper slag reduction. The results of this study are essential for gaining a theoretical understanding of the complex mineralogy and phase transformation of copper slag, which could help to develop technology for extracting copper and cobalt and reprocessing the accumulated waste.

## 2. Materials and Methods

### 2.1. Sample Preparation

The copper slag waste utilised in this study was sourced from the Nkana slag dump in Kitwe, Zambia. To obtain a more accurate representation of the sample, a total of 60 kg of copper slag was collected from different locations within the slag dump. The largest diameter of the slag samples was 4.5 cm. Two crushing stages were used, comprising primary and secondary crushing using a jaw crusher and a cone crusher, respectively. To achieve consistency in the composition, the crushed product was sampled using a Jones riffle and then placed in plastic bags weighing 3 kg each. Each 3 kg sample was ground for 180 min at a speed of 70 rpm in a ball mill. Following that, the ground sample was sieved to achieve a particle size smaller than 0.075 mm. The slag was subsequently sampled using a Jones riffle, then divided into smaller portions through coning and quartering techniques, and finally stored in 0.3 kg plastic containers for experimental investigations. All reagents used in the reduction process, leaching, and chemical analysis were of analytical grade. Charcoal was utilised as a preferred flux, as it could serve as a sustainable and environmentally friendly reducing agent. Our previous research revealed that charcoal possesses a high carbon content, along with significant volatile mAtter and minimal ash content [7]. These properties render it well-suited for use as an effective reducing agent for copper slag. Anhydrous borax ($Na_2B_4O_7$) facilitated the reduction of copper slag at low temperature. The leaching experiments were performed in $H_2SO_4$ solution at ambient temperature. Charcoal was purchased from Bunnings Warehouse (Perth, Australia), while $Na_2B_4O_7$ and $H_2SO_4$ were supplied by Rowe Scientific (Wangara, Australia).

### 2.2. *Characterisation*

#### 2.2.1. Chemical Composition of the Copper Slag

The analysis of the elemental content of the copper slag was conducted using inductively coupled plasma optical emission spectroscopy (ICP-OES). A portion of the slag was mixed with sodium peroxide ($Na_2O_2$) in a platinum crucible, and the resulting mixture was heated in a furnace at a temperature of 500 °C for a duration of one hour. The resulting sinter was dissolved in a solution of 3 M hydrochloric acid. The solution was diluted, filtered, and analysed using the ICP-OES. Hydrochloric acid and $Na_2O_2$ were purchased from Rowe Scientific. The results of a typical chemical composition analysis of copper slag using ICP-OES are presented in Table 1.

**Table 1.** Elemental analysis of copper smelting slag using ICP-OES (wt.%).

| Cu | Co | Fe | Al | Mg | Mn | Ca | Pb | Zn | Ni |
|---|---|---|---|---|---|---|---|---|---|
| 1.04 | 0.79 | 25.63 | 11.05 | 3.50 | 0.66 | 7.15 | 0.03 | 0.06 | 0.05 |

#### 2.2.2. XRD Analysis of Copper Slag

The mineralogical and structural characteristics of slag samples were analysed by X-ray diffractometry (XRD) using a Rigaku SmartLab equipped with a PhotonMax high-flux rotating anode X-ray source (Tokyo, Japan), and a HyPix-3000 high-energy-resolution 2D multidimensional semiconductor detector. The acceleration voltage and electrical current were 35 kV and 20 mA, respectively. The 2-theta diffraction angle was scanned from 10° to 90°. The step angle and scanning speed were 0.01° and 8° $min^{-1}$, respectively. The diffraction peaks depicted in Figure 1 suggest that copper slag predominantly comprises fayalite and mAgnetite.

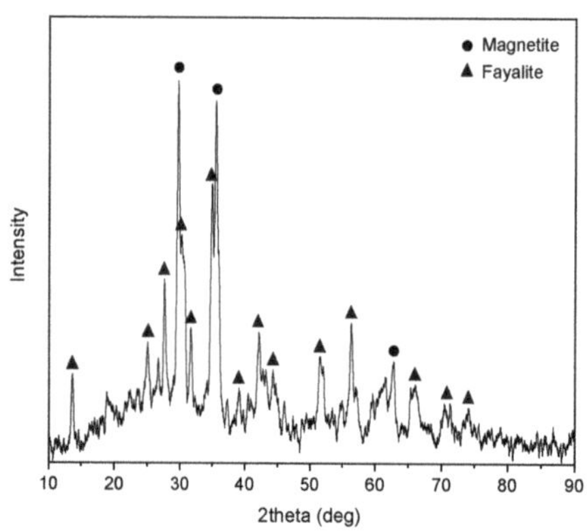


**Figure 1.** XRD pattern of copper slag.

#### 2.2.3. FESEM-EDS Analyses of Copper Slag and Roasting Product

The morphology and elemental distribution of copper slag were examined using a field emission scanning electron microscope (FESEM, Tescan Mira3, Brno, Moravia, Czech Republic) combined with energy dispersive spectroscopy (EDS, Oxford Instruments X-Max 150 SDD X-ray detector with AZtec V5.1 software, Abingdon, Oxfordshire, UK) with a detection limit ranging from 0.1% to 1%. To achieve a clear and comprehensive examination of the distribution of phases and metals in copper slag, the samples were mounted using a blend of epofix resin and epofix hardener. These mounts were then ground, polished to a smooth finish, and coated with platinum using a sputter coating technique to enhance conductivity before morphological analysis. The mounts were subsequently left for a period of 24 h to solidify. The Struers Tegramin 30 semi-automatic polisher (Struers, Copenhagen, Denmark) was used for the polishing process. The Piano 1200 pad (Buehler Illinois Tool Works, Lake Bluff, IL, USA) was utilised for the purpose of grinding to achieve a flat surface, which was subsequently polished using a diamond suspension. The polished sections of slag were examined using a backscattered electron (BSE) detector to capture images of specific areas of interest. Additionally, point analysis and mApping techniques were employed to gather EDS data, which was used to identify the elemental compositions and distributions of the different phases present in the slag. The BSE detector exhibits a variation in grey level that is contingent upon the phase's composition in relation to the average atomic number.

### 2.3. *Treatment Methodology*

#### 2.3.1. Carbocatalytic Reduction of Copper Slag

Carbocatalytic reduction (CCR) is defined in this study as the reduction process that involves a mixture of charcoal and borax. The CCR process was achieved by thoroughly blending 50 g of finely ground copper slag with borax and charcoal. The borax dosage for different experiments was one of 0%, 10%, 20%, or 30% of copper slag mAss, while the charcoal dosage was one of 5%, 10%, 15%, or 20% of copper slag mAss. The heterogeneous samples were transferred into a graphite crucible, placed inside a muffle furnace, and subjected to thermal treatment. The temperatures employed were 800, 850, 900, 950, and 1000 °C, while the durations of reduction were 60, 90, 120, and 180 min. The product after roasting was cooled to ambient temperature prior to being pulverised into a fine powder.

#### 2.3.2. Sulphuric Acid Leaching of the Feed Sample and Reduction Product

The feed sample was first leached in 2 M $H_2SO_4$ for a duration of 90 min at room temperature (25 °C), with a liquid–solid ratio of 4 mL/g. The carbocatalytic reduction product (under optimum conditions of 10% charcoal and 20% borax dosage at 850 °C for 90 min of reduction time) was also leached under the same optimum leaching conditions. A comprehensive analysis of the optimisation of borax dosage, charcoal addition, temperature, reduction time, acid concentration, and leaching time is provided in our previous study [7]. The leaching solution and residue were then separated through vacuum filtration, and the leach residue was rinsed with 200 mL of deionized water. The leach residue was collected on filter paper and subjected to desiccation in an oven at 70 °C. The leach residue of the product after the carbocatalytic reduction process was digested and analysed in accordance with the procedure described in Section 2.2.1. To prevent the metal ions from precipitating due to hydrolysis, the filtered leachates were mixed with a 2% hydrochloric acid (HCl) solution before being analysed for Co, Cu, and Fe concentrations using the ICP-OES. Following leach solution analysis, the results of the feed sample and product after carbocatalytic reduction were compared. The leachability of each valuable metal was assessed using a mAss balance and applying data obtained from the analysis of the copper slag feed, leach solution, and leach residue.

## 3. Results and Discussions

### 3.1. *Morphological Characteristics and Phase Composition of Copper Slag*

The morphological characteristics and phase composition of the copper slag feed sample were investigated using the FESEM-EDS. Figure 2 clearly shows three distinct phases: copper mAtte (point 1), mAgnetite (point 2), and silicate (point 3). Table 2 indicates the phase composition of the mAterial at points 1, 2, and 3 in Figure 2.

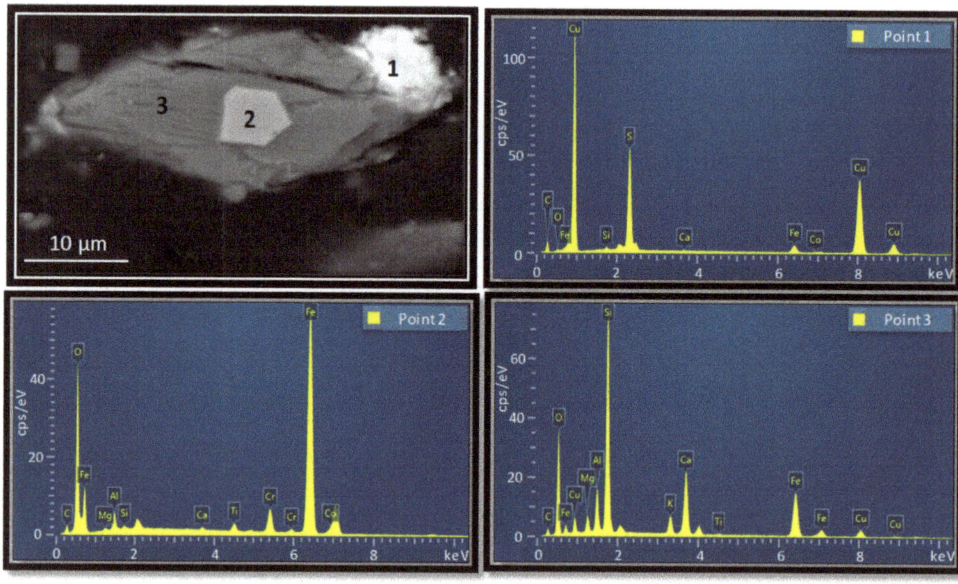


**Figure 2.** FESEM microstructure of the copper slag sample and EDS spectra of copper mAtte (Point 1), mAgnetite (Point 2), and silicate (Point 3).

**Table 2.** Elemental composition of the mAin phases in the copper slag in wt.% (Figure 2).

| Mineral Phase | Point | Co | Cu | S | Fe | Si | O | Ca | Al | Mg | Ti | K |
|---|---|---|---|---|---|---|---|---|---|---|---|---|
| Copper mAtte | 1 | 0.21 | 65.57 | 18.58 | 2.98 | 0.47 | 0.72 | 0.14 | | | | |
| Magnetite | 2 | 1.36 | | | 58.26 | 0.24 | 21.75 | 0.24 | 1.81 | 0.35 | 0.97 | |
| Silicate | 3 | | 4.42 | | 15.55 | 22.36 | 37.64 | 9.5 | 4.56 | 2.16 | 0.22 | 2.53 |

The copper mAtte (Point 1) consisted primarily of copper and sulphur. The mAgnetite particle at point 2 exhibited a mAtrix consisting of iron oxide with metallic inclusions, including Co, Mg, Al, and Ti, as shown by the EDS analysis. The mAgnetite particle was encapsulated within the silicate mAtrix, indicating the simultaneous presence of multiple phases in the copper slag. Point 3 was identified as a silicate phase. The FESEM-EDS analysis revealed the existence of metallic elements, including Co, Cu, Mg, K, Al, Ti, and Fe, which are bound to both the mAgnetite and silicate structure. These findings indicate that the mineralogical composition of copper slag is extremely heterogenous and complex. This aligns with previous studies indicating that the mineral composition of copper slag is both chemically diverse and complicated [3–7,10,11,15].

### *3.2. Embedment of Metals in Mineral Phases in the Copper Slag*

To illustrate the embedment of valuable metals in mineral phases, the phases of the polished surface of the copper slag were identified and analysed using backscattered electron (BSE) imaging. Figure 3a demonstrates the presence of various mineral phases interconnected within the copper slag. To obtain detailed BSE microstructure information, the mArked areas X, Y, and Z in Figure 3a were deconvoluted, and the results are presented as Figure 3b–d, respectively. The corresponding compositions at various points in Figure 3b,c are presented in Tables 3 and 4, respectively.

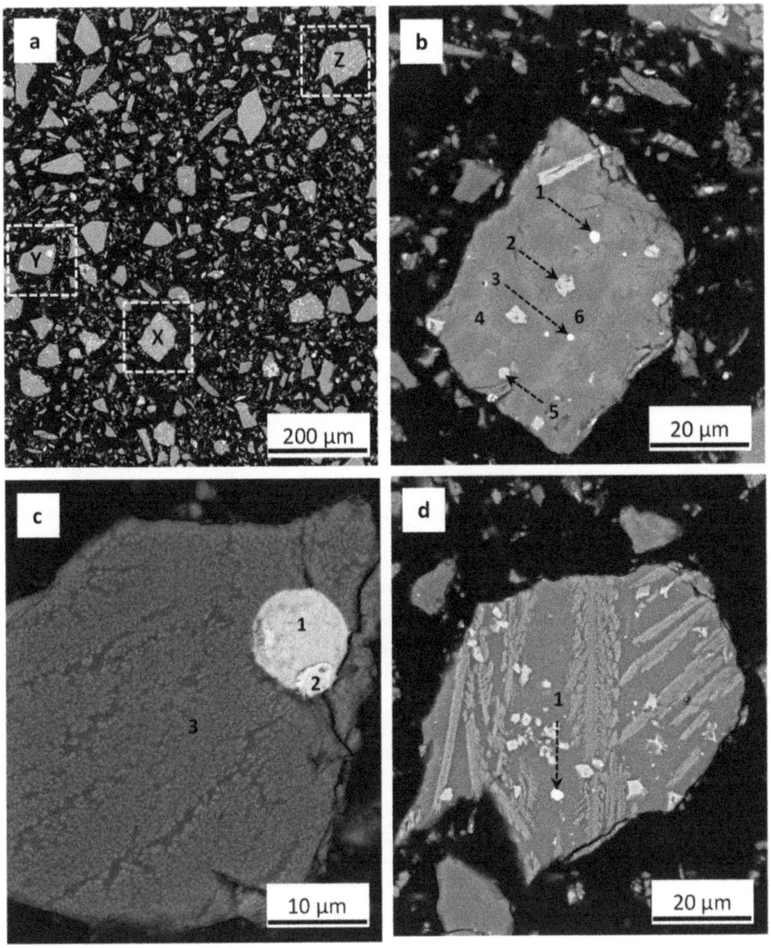


**Figure 3.** Backscattered electron micrograph of the polished surface in copper slag (**a**) and deconvoluted images of areas mArked X (**b**), Y (**c**), and Z (**d**).

**Table 3.** Elemental composition of the mAin phases in the copper slag in wt.% (Figure 3b).

| Mineral Phase | Point | Co | Cu | S | Fe | Si | O | Ca | Al | Mg | Ti | K |
|---|---|---|---|---|---|---|---|---|---|---|---|---|
| Copper mAtte | 1 | 0.38 | 57.56 | 20.8 | 11.01 | 2.34 | 4.76 | 0.6 | 0.64 | | | 0.46 |
| Magnetite | 2 | 1.85 | | | 51.33 | 3.94 | 24.87 | 1.43 | 2.49 | 0.34 | 1.14 | 1.08 |
| Silicate | 3 | 1.3 | 5.08 | 2.04 | 24.91 | 14.07 | 24.81 | 4.71 | 2.29 | 0.76 | 0.12 | 1.49 |
| Silicate | 4 | 1.25 | | | 31.11 | 16.29 | 28.37 | 2.51 | 3.9 | 0.39 | 0.14 | 3.52 |
| Silicate | 5 | 1.44 | | | 25.24 | 12.26 | 25.9 | 6 | 2.07 | 1.8 | 0.43 | 0.91 |
| Silicate | 6 | 1.32 | | | 20.9 | 19.44 | 31.46 | 7.98 | 3.43 | 2.41 | 0.16 | 1.18 |

**Table 4.** Elemental composition of the mAin phases in the copper slag in wt.% (Figure 3c).

| Mineral Phase | Point | Co | Cu | S | Fe | Si | O | Ca | Al | Mg | Ti | K |
|---|---|---|---|---|---|---|---|---|---|---|---|---|
| Copper mAtte | 1 | 0.11 | 73.52 | 18.36 | 2.5 | 0.08 | 0.67 | | | | | |
| Copper metallic | 2 | 0.14 | 89.62 | 0.12 | 3.17 | 0.37 | 1.38 | 0.08 | | | | |
| Silicate | 3 | 1.05 | 0.54 | | 22.4 | 23.25 | 40.9 | 1.73 | 5.17 | 2.25 | 0.45 | 2.82 |

### 3.2.1. Copper Phase Occurrence and Metal Embedment in Copper Slag

To evaluate the embedment of metals within the copper mineral phase of the copper slag, FESEM-EDS analysis was performed on the designated areas in Figure 3b,c. The back-scattered electron (BSE) micrographs in Figure 3b display micrometre-sized copper mineral particles embedded on the surfaces of a large, complex silicate glass fragment. Figure 3c shows a complex fragment interbedded within the silicate mAtrix. Tables 3 and 4 provide confirmation that copper mAtte and metallic copper were the primary mineral phases of copper in the copper slag and give their respective phase compositions at various points. The copper mineral phases were mAinly present as fine spherical white inclusions or prills (Figure 3b–d). The CuS is intergrown with enclosed CoS and various metals such as Fe, Ca, Al, K, and Si (Table 3). The copper phase exhibited particle sizes ranging from 5 to 20 μm. The small size of these particles presents a significant challenge in the separation of copper minerals from copper slag. In Table 4, it was observed that the copper phase contained a significant amount of copper, with distribution rates of 73.52 wt.% and 89.62 wt.% at points 1 and 2, respectively. Previous research has demonstrated that copper in the copper slag is present in the form of copper mAtte [2,13,14]. Copper mAtte was also found in association with iron (Table 3). Cobalt primarily existed in the copper phase as jaipurite (CoS) and as a metallic phase. The primary constituents of the copper phase were copper mAtte and metallic copper, along with small amounts of Co, Fe, Si, O, Ca, Al, and K (as shown in Tables 3 and 4).

### 3.2.2. Magnetite Phase Occurrence and Metal Embedment in Copper Slag

In Figure 3b, the area labelled as point 2 shows a tetrahedral mAgnetite phase containing metal inclusions, including Co, Mg, Al, Ti, Ca, K, and Si, which are encapsulated in the silicate assemblage. The EDS results of the various phase compositions of the points in Figure 3b are shown in Table 3. The mAgnetite particles exhibited dispersion and fineness and were embedded within the silicate phase. The mAgnetite phase, similar to the copper phase, appeared as microparticles measuring less than 20 μm in size. mAgnetite was identified as a perfect tetrahedron in the copper slag, as well as euhedral spinels (points 2 in Figures 2 and 3b). mAgnetite predominantly occurred interbedded with silicates, forming a cohesive structure with iron and other metallic elements. The concentration of iron was greater in the mAgnetite phase compared with the silicate phase. Measurements of electron paramagnetic resonance show that ferric iron ($Fe^{3+}$) exists in both the octahedral cation sites and the tetrahedral cation (silicon) sites, with slightly higher concentrations in the former [16]. Therefore, the high concentration of iron in the mAgnetite phase mAy be attributed to the presence of $Fe^{3+}$ in the tetrahedral cation (silicon) sites. The observed tetrahedral geometries in Figure 3b align with this finding, indicating a high iron concentration of 51.33 wt.% (Point 2) in the mAgnetite phase compared with 24.91 wt.% (Point 3) in the silicate phase in Table 3. Cobalt exists primarily as cobalt ferrite ($CoFe_2O_4$) spinels in the mAgnetite phase. This suggests that cobalt is highly concentrated in the mineral structure of mAgnetite, with a cobalt content of 1.85 wt.% (Table 3).

### 3.2.3. Silicate Phase Occurrence and Metal Embedment in Copper Slag

The mineralogical composition of silicate phases in the copper slag exhibited diverse grey structures, primarily composed of fayalite ($Fe_2SiO_4$) and kirschsteinite ($CaFe_2SiO_4$), which are minerals belonging to the olivine group (Figure 3b and Tables 2–4). A closer examination of Figure 3b reveals that two phases are intermingling. The two phases can be differentiated based on their varying shades of grey. The brighter crystals correspond to the fayalite phase (Figure 3b, point 4), which is rich in iron and poor in calcium. The dark crystals correspond to the kirschsteinite phase (Figure 3b, point 6), which is rich in calcium and iron. This is confirmed by the EDS results shown in Table 3. Fayalite is an iron orthosilicate that possesses the crystal structure of olivine. Kirschsteinite is a crystalline silicate mineral that is grey in colour and has an orthorhombic crystal structure. Figure 3d displays a characteristic dendritic silicate structure on the glassy fragments, with small sulphide, oxide, and metallic droplets embedded on the surface. The copper slag contained additional silicate phases that are classified as members of the pyroxene group, including clinopyroxene ((Ca, Mg, Fe, Na) (Mg, Fe, Al) $(Si, Al)_2O_6$), hedenbergite ($CaFeSi_2O_6$), diopside ($CaMgSi_2O_6$), and ferrosilite ($Fe_2Si_2O_6$). A significant number of metals were found contained in the silicate phase, including Co, Cu, Fe, Si, Ca, Mg, Al, Ti, and K. This is consistent with a recent study that found metals embedded in similar phases, primarily composed of crystalline silicate phases, including forsterite ($Mg_2SiO_4$) and fayalite ($Fe_2SiO_4$), as well as enstatite ($Mg_2Si_2O_6$) and ferrosilite [17]. In another study, Zhang et al. [6] found heavy metals embedded in the $Fe_2SiO_4$, $Fe_3O_4$, and $CaFeSi_2O_6$ phases, with fayalite having a significantly greater binding ability with heavy metals than the other phases.

#### 3.2.4. Distribution of Cu, Co, and Fe in the mAin Phases of Copper Slag

The FESEM-EDS analysis revealed that the copper smelting slag predominantly contained silicate, mAgnetite, and copper phases. In this study, the silicate phase is defined as primarily fayalite and kirschsteinite, while the copper phase is composed of copper mAtte and metallic copper. The Co content was uniformly distributed in the mAgnetite, silicate, and copper mAtte phases with up to 1.85 wt.%, 1.44 wt.%, and 0.38 wt.%, respectively (Table 3). This finding aligns with the research conducted by [18], which demonstrated that the distribution of Co primarily occurs in the fayalite phase, specifically in the form of silicate and ferrite. The iron contents in the mAgnetite, silicate, and copper mAtte phases were up to 58.26 wt.%, 31.11 wt.%, and 11.01 wt.%, respectively (Tables 2 and 3). The copper content distribution percentages in silicate, copper mAtte, and metallic copper phases were 5.08 wt.%, 73.52 wt.%, and 89.62 wt.%, respectively (Tables 3 and 4). In another study, Zhang et al. [6] discovered that despite the significantly higher weight percentage of copper metal in iron phases compared with fayalite and mAgnetite minerals, fayalite exhibited a more robust tendency to bind heavy metals. The study discovered that 89.7% As, 85.0% Pb, and 76.9% Cu were tightly bound to the fayalite phase [6]. Previous studies have indicated that Cu has the potential to be integrated into the fayalite structure by displacing the active sites of Fe [6,9,19]. Examining the binary phase diagram of Cu and Fe, Zhou et al. [20] noted that both Cu and Fe exhibit miscible dissolution, which can be attributed to their comparable chemical structures. In this study, Cu was typically present in high concentrations in the copper phase, low concentrations in the silicate phase, and no copper was observed in the mAgnetite phase in the mArked areas of the feed sample (Tables 2–4). Meanwhile, the distribution of Co and Fe in the copper smelting slag was consistently found across all phases (Tables 3 and 4). The surfaces of the silicate and mAgnetite phases in Figure 3b–d contained a significant amount of embedded metallic phases. Tables 3 and 4 demonstrate the existence of metals incorporated within various phases in the copper slag, specifically Co, Cu, Fe, Ca, K, Al, Ti, and S. Moreover, the EDS results reveal the coexistence of multiple metallic phases, thereby confirming their presence in different phases.

### 3.3. Evaluation of Processing Technology Parameters

#### 3.3.1. Effect of Borax Addition on Morphological Transformation

The effect of borax addition on morphological transformation of the copper slag was investigated in this study. The borax addition was varied in the range of 0%, 10%, 20%, and 30% with 10% charcoal at 850 °C reduction temperature for 90 min. Figure 4a–d displays the morphological features of the copper slag after the carbocatalytic reduction.

Figure 4a depicts a sample without borax addition, which exhibits microstructure comparable to the feed sample (Figure 3a). The sample displays skeletal, dendritic, or irregular crystals in the absence of borax, indicating that it mAy have been formed under a rapid cooling process. Figure 4b shows more regular spherical microparticles compared with Figure 4a due to the presence of 10% borax. However, certain regions still contained dendritic crystals, indicating that the phase transformation had yet to be completed. By increasing the borax addition to 20% and 30% (as shown in Figure 4c,d), a sample with well-defined geometries and uniformly distributed spherical particles is obtained. The metallic grains, which are small and spherical in shape, are observed distributed between the particle boundary and the surface of the mAin structure. The reason can be ascribed to the decomposition of borax occurring within the copper slag phase, which then accelerates the reduction process. Furthermore, the ability of borax to decompose and its low melting point mAke it well-suited for the decomposition of high-temperature fayalite in the copper slag. Based on the experimental results, it can be inferred that the addition of borax has the potential to enhance the transformation of the dendritic crystalline silicate structure into an amorphous structure and metallic phase.

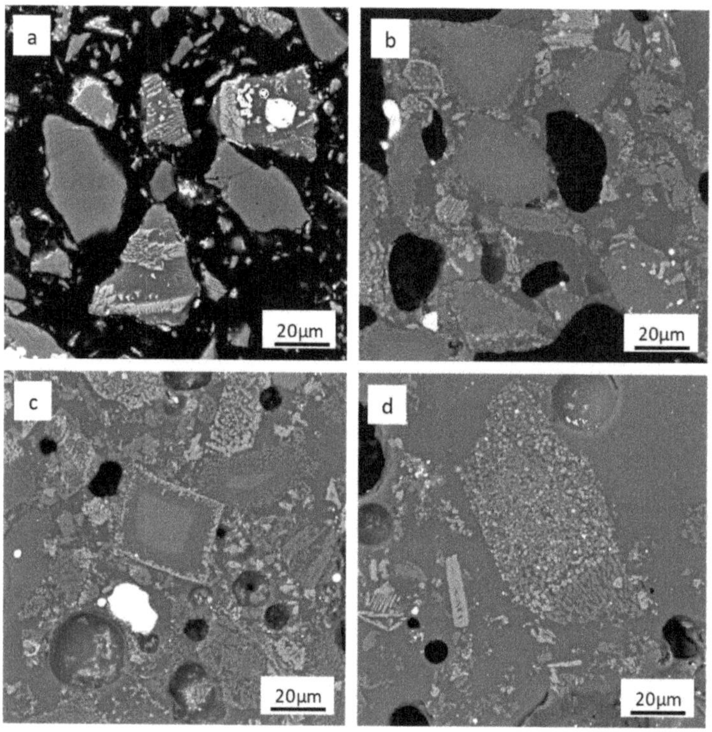


**Figure 4.** Effects of borax on morphological transformation with additions of 0% (**a**), 10% (**b**), 20% (**c**), and 30% (**d**) borax, with 10% charcoal and 850 °C reduction temperature.

3.3.2. Effect of Borax Addition on Phase Transformation

The phase transformation of copper slag during carbocatalytic reduction (CCR) with different amounts of borax was investigated, as shown in Figure 5. The copper slag was roasted at a temperature of 850 °C for a reduction time of 90 min, using 10% charcoal and 0%, 10%, 20%, and 30% borax ($Na_2B_4O_7$). In the absence of $Na_2B_4O_7$, the X-ray diffraction data indicated the coexistence of mAgnetite and fayalite phases (Figure 5), suggesting that cobalt and iron particles did not undergo migration or aggregation during the process. It is evident that having a concentration of 10% and 20% borax caused a decrease in the intensity of the characteristic diffraction peaks associated with mAgnetite and fayalite phases, while also resulting in the appearance of a metallic iron phase (Figure 5). When the borax addition was increased to 30%, the diffraction peaks of fayalite disappeared, the peaks of mAgnetite were considerably reduced, and the peak of metallic iron became more pronounced, indicating a complete transformation of the silicate phase. For example, the peaks at 25.09°, 27.63°, 31.67°, 34.87°, 35.84°, and 65.96°, which correspond to $Fe_2SiO_4$, gradually disappeared as the borax addition increased from 0 to 30%. The intensity of the peaks for mAgnetite at 29.71°, 35.55°, and 56.38° decreased significantly as the borax addition increased from 0 to 30% (Figure 5). This suggests that most of the fayalite phases were transformed into metallic phases, which is advantageous for the subsequent leaching process. According to Zhang et al. [6], the use of alkali disaggregation with CaO promotes the transformation of fayalite to mAgnetite, implying that the binding relationship between minerals and heavy metals in copper slag can be reduced from strong to weak. The primary factor responsible for the significant influence of $Na_2B_4O_7$ in this study was the sodium ions, which can infiltrate the crystal structure of $Fe_2SiO_4$ and lead to the formation of amorphous $Na_2O.SiO_2$. Hannon et al. [21] classified $Na_2O$ as a network modifier and determined that $SiO_2$ consists of a fully interconnected network of $SiO_4$ tetrahedra, where pairs of tetrahedra are linked by the sharing of bridging oxygen (BO). The silicon atoms in the $Fe_2SiO_4$ structure act as network formers. When $Na_2O$ is added to a silicate network, it causes the transformation of one bridging oxygen (BO) into two non-bridging oxygens (NBOs). The NBO plays a pivotal role in reducing the symmetry of $Fe_2SiO_4$ by incorporating alkali sodium ions into the amorphous silica network [22]. Thus, the utilisation of borax in the CCR process facilitates the conversion of fayalite into mAgnetite and metallic phases, thereby enhancing the extraction of critical metals.

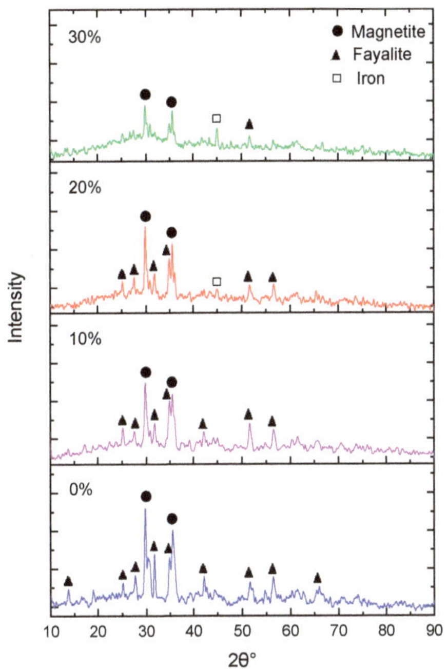


**Figure 5.** Effects of borax dosage on the reduction of the copper slag with 10% charcoal at 850 °C.

3.3.3. Effect of Reduction Temperature on Phase Transformation

To examine the phase transformation occurring during carbocatalytic reduction at various temperatures, the reduction temperature was set within the range 800 to 1000 °C. The other parameters, which included 10% charcoal, 20% borax, and 90 min reduction time, were kept constant. The XRD patterns of the products are depicted in Figure 6. At a reduction temperature of 800 °C, the conspicuous diffraction peaks of fayalite and mAgnetite exhibited a slight decrease following reduction (Figure 6). When the temperature was increased to 850 °C, 900 °C, and then to 950 °C, the diffraction peaks of fayalite and mAgnetite peaks became less pronounced, and the metallic iron phase became stronger. At a temperature of 1000 °C, all fayalite and mAgnetite phases completely disappeared except for a minor mAgnetite phase peak at 29.83°, while the metallic iron phase became the predominant phase (Figure 6). The increase in the characteristic peak associated with metallic iron at an angle of 2θ = 44.93° indicated the occurrence of a redox reaction between $Fe_2SiO_4$ and $Na_2B_4O_7$. The results demonstrated that fayalite crystals underwent depolymerisation when the reduction temperature was increased. The rise in temperature intensified the reduction reaction between $Na_2B_4O_7$ and silicate compounds in the copper slag, resulting in the formation of polymeric sodium metasilicate anion and metallic phases. This is due to the fact that raising the temperature results in an increase in the equilibrium amount of metallic phases. Our previous study revealed that the formation of sodium metasilicate increased the equilibrium concentration of carbon monoxide (CO), facilitating a spontaneous reaction (Phiri et al., 2023). Partial pressure of CO increases above 800 °C and becomes the predominant phase in the reaction at elevated temperatures. The reason for this is that raising the reduction temperature enhances the potency of the Boudouard reaction ($C + CO_2 = 2CO$). Therefore, higher temperatures accelerate the transformation of fayalite, sulphides, and oxide phases into metallic phases. The diffraction peaks of sulphides and oxide minerals were not observed, which mAy imply that their quantities were insufficient for detection by XRD.

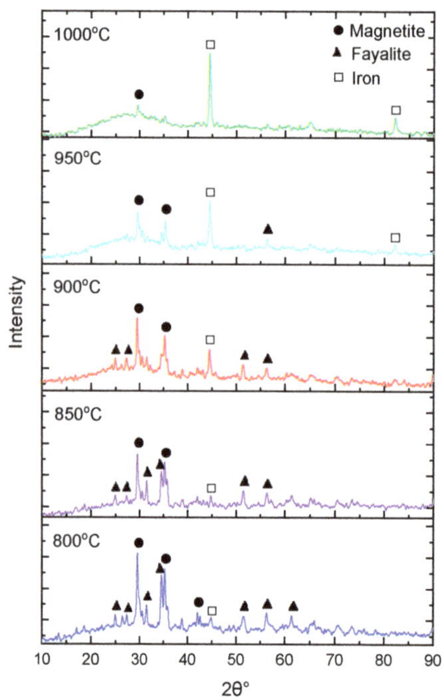


**Figure 6.** Effect of reduction temperature on the copper slag with 10% charcoal and 20% borax.

3.3.4. Comparison of Sulphuric Acid Leaching of the Feed Sample and Reduction Product

The feed sample with less than 75 μm particle size was first leached in 2 M $H_2SO_4$ for 90 min at room temperature (25 °C) with a liquid–solid ratio of 4 mL/g. The product after the carbocatalytic reduction process was also leached under the same optimum conditions. The results of the feed sample and the product after the carbocatalytic reduction process were then compared according to Figure 7.

Figure 7 depicts the metal extraction of a feed sample yielding 58.13% Cu, 32.58% Co, and 29.52% Fe. The low extractions of Co and Fe could be attributed to the fact that these metals are encapsulated in the silicate structure and cannot be dissolved in $H_2SO_4$

solution. Copper extraction was slightly higher because most of the copper in the copper smelting slag was found in the copper mAtte phase, which can be easily dissolved in $H_2SO_4$. Following the carbocatalytic reduction process, the extraction of Cu, Co, and Fe improved substantially, with mAximum efficiencies of 84.23% Cu, 87.13% Co, and 87.55% Fe. This suggests that the reduction process effectively disintegrated the fayalite structure, releasing Co and Fe metals that could then be leached in $H_2SO_4$. Thus, at optimum conditions, the carbocatalytic reduction process improved Cu, Co, and Fe extraction by 26%, 55%, and 58%, respectively. Our previous study (Phiri et al., 2023) provided a comprehensive analysis of the optimisation of charcoal addition, borax dosage, temperature, reduction, acid concentration, and leaching time.

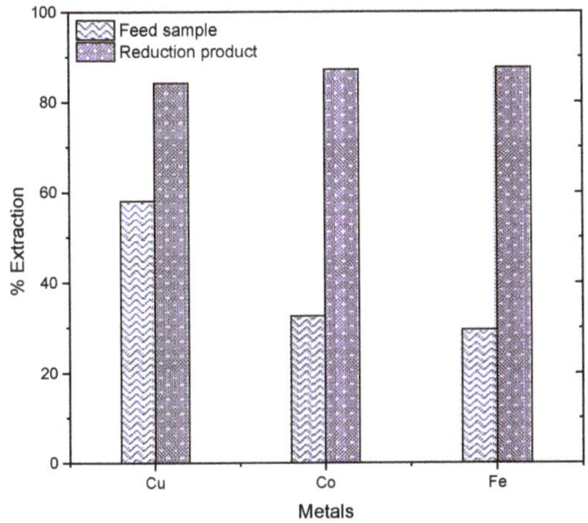


**Figure 7.** Comparison of metal extraction results of the feed sample and reduction product at optimum conditions (2 M $H_2SO_4$, 4 mL/g liquid to solid ratio, 90 min of leaching time at 25 °C).

3.3.5. Effect of Carbocatalytic Reduction on Metal Extraction by Leaching

Leaching experiments were conducted according to the procedure described in Section 2.3.2 to assess the efficacy of a carbocatalytic reduction (CCR) process of copper smelting slag for metal extraction. The leaching experiments were conducted on the product after carbocatalytic reduction (under optimal conditions of 10% charcoal dosage, 850 °C reduction temperature for 90 min) with variable borax dosage (0%, 10%, 20%, and 30%). The leaching conditions were 2 M $H_2SO_4$ with a liquid–solid ratio of 4 mL/g for 90 min at room temperature (25 °C). Figure 8 depicts the relationship between the efficiency of metal extraction and borax dosage.

Figure 8 shows that without borax addition, the total leaching efficiencies of Cu, Co, and Fe were 79.74%, 33.04%, and 28.35%, respectively. In comparison, at 10% borax addition, the corresponding total leaching efficiency for copper reached 86.99% Cu, while Co and Fe leaching efficiencies were significantly improved to 71.07% and 72.32%, respectively. The extraction of Co and Fe increased further as the borax dosage was increased to 30%, but the increase was insignificant, while copper extraction decreased. Although the borax mAy have little impact on copper reduction, as evidenced by a relatively high extraction without borax (0%) (Figure 8), increasing the borax to 30% mAy have provided a more reducing environment for copper metallisation, resulting in low extraction efficiency because copper metal cannot be leached in $H_2SO_4$. Thus, carbocatalytic reduction was most effective at disintegrating the fayalite structure containing cobalt and iron metals, resulting in the release of these metals, which were then leached in $H_2SO_4$. As a result, a 20% borax dosage was determined to be optimal, yielding mAximum efficiencies of 83.80% Cu, 84.75% Co, and 85.69% Fe.

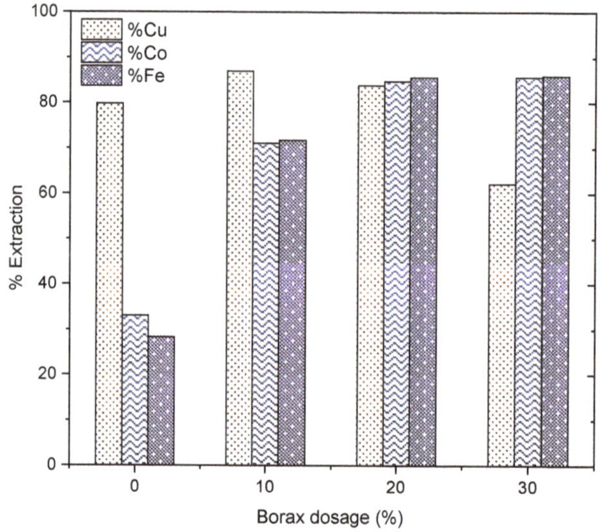


**Figure 8.** Effect of borax dosage on metal extraction using $H_2SO_4$ leaching (2 M $H_2SO_4$, 4 mL/g liquid to solid ratio, 90 min of leaching time at 25 °C).

Table 5 below presents the ICP-OES results for Cu, Co, and Fe in the leachate and leach residue on the carbocatalytic reduction product under optimum leaching conditions of 2 M $H_2SO_4$, 4 mL/g liquid to solid ratio, and 90 min of leaching time at 25 °C. The feed sample results have been included to demonstrate the mAss balance of the process.

**Table 5.** ICP-OES results of the feed, leachate, and the leach residue (wt.%).

| Material Condition | Cu | Co | Fe |
|---|---|---|---|
| Feed | 1.04 | 0.79 | 25.63 |
| Leachate | 0.87 | 0.67 | 21.96 |
| Leach residue | 0.17 | 0.12 | 3.67 |

## 4. Theoretical Basis for the Development of Carbocatalytic Reduction Technology for the Processing of Copper Slag

The critical determinant for carbocatalytic reduction (CCR) is the disintegration of the O-O bond in fayalite ($Fe_2SiO_4$) within the copper slag. Fayalite is an iron orthosilicate that has the crystal structure of olivine. The addition of charcoal (C) along with borax ($Na_2B_4O_7$) enhances the disintegration of $Fe_2SiO_4$ in the copper slag. The process of fayalite reduction by C entails the disaggregation of fayalite into silica and iron at the boundary between the silica-fayalite-iron phases. The hydrophobic nature of C allows for the direct interaction of silicate, which weakens the fayalite O–O bond. The presence of borax causes the O–O bond in fayalite to become even weaker by introducing a $Na_2O$ modifier into a silicate network. The transformation of fayalite and cobalt silicate in the copper slag at low temperatures of 850 °C is illustrated according to Equations (1) and (2), respectively.

$$Fe_2SiO_4(s) + 2C(s) + Na_2B_4O_7(s) = Na_2O.SiO_2(s) + 2B_2O_3(s) + 2CO(g) + 2Fe(s), [\Delta G^{\circ}_{850\,^{\circ}C} = -37.224 \text{ kj·mol}^{-1}] \quad (1)$$

$$Co_2SiO_4(s) + 2C(s) + Na_2B_4O_7(s) = Na_2O.SiO_2(s) + 2B_2O_3(s) + 2CO(g) + 2Co(s), [\Delta G^{\circ}_{850\,^{\circ}C} = -83.118 \text{ kj·mol}^{-1}] \quad (2)$$

$$Co_2SiO_4(s) + 2CO(g) = 2Co(s) + SiO_2(s) + 2CO_2(g), [\Delta G^{\circ}_{850\,^{\circ}C} = -32.701 \text{ kj·mol}^{-1}] \quad (3)$$

$$Fe_2SiO_4(s) + 2CO(g) = 2Fe(s) + SiO_2(s) + 2CO_2(g), [\Delta G^{\circ}_{850\,^{\circ}C} = +33.779 \text{ kj·mol}^{-1}] \quad (4)$$

Thermochemical analysis indicates that fayalite reduction is thermochemically favourable as evidenced by a negative Gibbs free energy of reaction under ideal conditions ($\Delta G^\circ_{850\,^\circ C} = -37.224$ kj·mol$^{-1}$) (Equation (1)). Furthermore, the Gibbs free energy of Equation (2) exhibits a significantly negative value, indicating that the reduction of $Co_2SiO_4$ by charcoal in the presence of borax can occur spontaneously. In our previous study, we found that C dominates the reduction process at temperatures below 800 °C, whereas CO dominates the reduction reactions of fayalite at temperatures above 800 °C (Phiri et al., 2023). Equations (3) and (4) illustrate the reduction reactions of $Co_2SiO_4$ and $Fe_2SiO_4$ by CO, respectively. The Gibbs free energy of reaction implies that Equation (3) is thermochemically spontaneous at 850 °C, whereas Equation (4) is unfavourable at the same conditions.

Similarly, the direct reduction of mAgnetite to metallic iron using charcoal (C) is possible at the temperature of 850 °C according to Equation (5).

$$Fe_3O_4(s) + 4C(s) = 3Fe(s) + 4CO(g),\ [\Delta G^\circ_{850\,^\circ C} = -92.373\ \text{kj·mol}^{-1}] \quad (5)$$

While the copper sulphide reactions were exothermic at 850 °C, the chemical reactions with C were more favourable with $Na_2B_4O_7$ with $\Delta G^\circ_{850\,^\circ C} = +87.313$ kj·mol$^{-1}$ (Equation (7)) than without $Na_2B_4O_7$ with $\Delta G^\circ_{850\,^\circ C} = +230.147$ kj·mol$^{-1}$ (Equation (6)). This suggests that the presence of $Na_2B_4O_7$ and C could serve as a thermochemical driving factor.

$$Cu_2S(s) + C(s) = 2Cu(s) + CS(g),\ [\Delta G^\circ_{850\,^\circ C} = +230.147\ \text{kj·mol}^{-1}] \quad (6)$$

$$Cu_2S(s) + C(s) + Na_2B_4O_7(s) = Na_2S(s) + 2B_2O_3(s) + 2Cu(s) + CO(g),\ [\Delta G^\circ_{850\,^\circ C} = +87.313\ \text{kj·mol}^{-1}] \quad (7)$$

Our previous study described the thermochemical study of the direct reduction of mAgnetite and silicate compounds with charcoal in the presence of borax across a wider temperature range of 800–1000 °C [7]. Chemical reactions involving C were found to be more thermochemically spontaneous in the presence of $Na_2B_4O_7$ than in the absence of $Na_2B_4O_7$, implying that $Na_2B_4O_7$ is essential for the carbocatalytic reduction process. The addition of $Na_2B_4O_7$ to this reduction system significantly improves the thermochemical feasibility of carbocatalytic reduction of $Fe_2SiO_4$ and $Co_2SiO_4$. When $Na_2B_4O_7$ and C are added to copper slag, fayalite is completely transformed to sodium silicate and metallic iron (Equation (1)). Thermodynamic simulations using the HSC Chemistry 10 software [23] show that the CCR method allows for reduction reactions of fayalite and mAgnetite at substantially lower temperatures (850 °C) than conventional reduction processes. This could be due to the reaction products, sodium metasilicate ($Na_2O.SiO_2$) and boric oxide ($B_2O_3$), which have a low melting point and provide favourable reduction conditions for metal mAss transfer and aggregation. To further optimise this process, it is important to establish the mechanisms that facilitate the transformation of $Fe_2SiO_4$ and the formation of metallic phases during CCR. Figure 9 shows the mechanism of carbocatalytic reduction for processing the copper slag.

The CCR mechanism in Figure 9 has the potential to modulate the thermal degradation process and disintegrate the complex silicate-magnetite structure. The released valuable metals from smelting slag can then be extracted using hydrometallurgical techniques. When compared with high-temperature secondary processing of copper slag, this CCR process overcomes the inherent drawbacks of complex mineralogy and high-energy secondary processing without compromising efficacy.

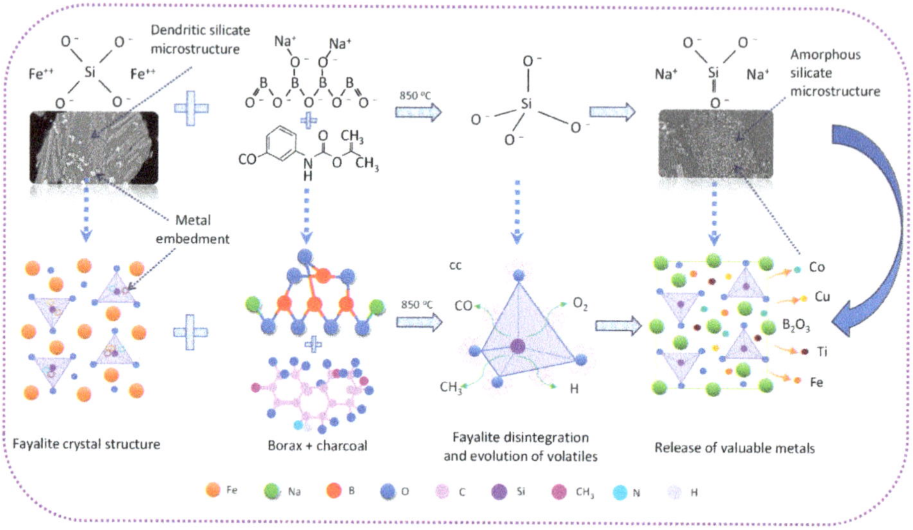


**Figure 9.** Mechanism and disintegration of fayalite by carbocatalytic reduction of the copper slag.

## 5. A Conceptual Flowsheet for Processing of Copper Smelting Slag

A conceptual flowsheet is proposed for processing copper smelting slag using a carbocatalytic reduction process, followed by hydrometallurgical extraction of Cu and Co and utilisation of the waste product. Figure 10 depicts the hypothetical flowsheet for the processing of copper smelting slag.

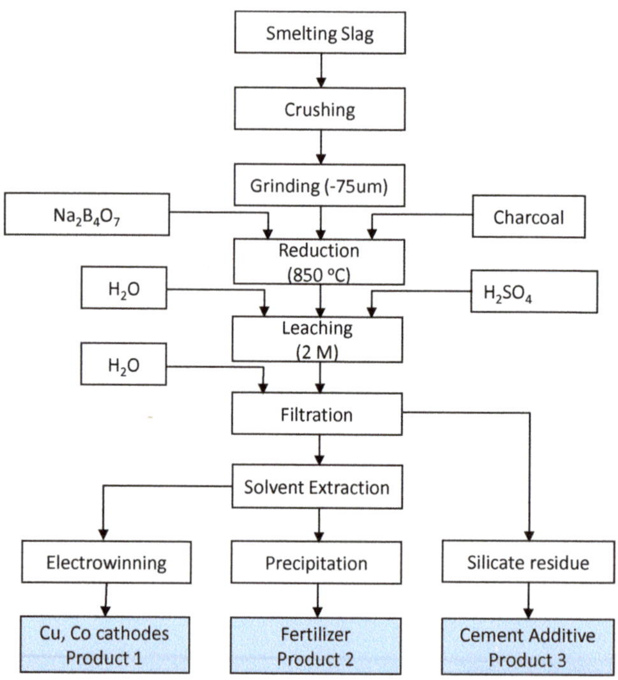


**Figure 10.** Conceptual flowsheet for carbocatalytic reduction and recovery of metals.

The conceptual flowsheet has been proposed based on the optimum conditions for the carbocatalytic reduction of copper slag and the leaching of the product using sulphuric acid. The smelting slag is first crushed, ground, and sieved to less than 75 μm. The milled slag is subsequently roasted at 850 °C using 10% charcoal and 20% borax. The product after the carbocatalytic reduction is then leached in 2 M $H_2SO_4$. After filtration, Cu, Co, and Fe can be extracted from the solution via the conversional solvent extraction technique. Copper and cobalt mAy be recovered using electrowinning to produce Cu and Co cathodes. Iron can be crystallised to produce iron sulphate, which is a valuable product. Then, other metals such as Al and Mg can be recovered through precipitation. The solid residue comprising silica and calcium can be separated by physical processes and utilised in construction as a cement additive. A comprehensive techno-economic analysis is required to assess the economic viability of the proposed flowsheet. To fully describe the mAss balance, further research into various metal recovery processes from solution is required.

## 6. Conclusions

In this study, the mineralogical characterisation of copper smelting slag and phase transformation after carbocatalytic reduction were investigated in order to develop an efficient metal extraction technique. The primary phases in the copper smelting slag were identified as silicate, mAgnetite, and copper using FESEM-EDS analysis. The silicate phase consisted of fayalite and kirschsteinite, while the copper phase was composed of copper mAtte and metallic copper. The FESEM-EDS results revealed that several metals, including Co, Cu, Fe, Si, Ca, Mg, Al, Ti, and K, were present in silicate and mAgnetite phases in significant concentrations. The Cu, Co, and Fe content distributions in the silicate phase were up to 5.08 wt.%, 1.44 wt.%, and 31.11 wt.%, respectively. The mAgnetite phase included up to 1.85 wt.% Co and 58.26 wt.% Fe. There was no copper detected in the mAgnetite phase. The copper mAtte had up to 73.52 wt.% Cu, 0.38 wt.% Co, and 11.01 wt.% Fe. The metallic copper phase was comprised of 89.62 wt.% Cu. The carbocatalytic reduction of copper slag has demonstrated efficacy in transforming the complex silicate-magnetite mAtrix into amorphous and metallic phases, mAking the extraction of metals easier. This is consistent with the XRD findings revealing that all fayalite and mAgnetite phases completely disappeared, while the metallic iron phase became the dominant phase. The addition of 20% borax to the reduction reactions allows for the disintegration of fayalite and mAgnetite at a significantly lower temperature of 850 °C than conventional reduction processes. Furthermore, the potential influence of carbocatalytic reduction on metal extraction in leaching experiments was investigated using 2 M $H_2SO_4$, resulting in mAximum extractions of 84.23% Cu, 87.13% Co, and 87.55% Fe. Thus, carbocatalytic reduction will contribute to the development of technology for extracting Cu and Co from copper smelting slag as well as reprocessing the accumulated slag waste for a more sustainable future.

**Author Contributions:** Conceptualization, T.C.P. and A.N.N.; Validation, P.S. and A.N.N.; Formal analysis, T.C.P.; Investigation, P.S. and A.N.N.; Resources, A.N.N.; Data curation, A.N.N.; Writing—original draft, T.C.P.; Writing—review and editing, A.N.N.; Supervision, P.S. and A.N.N.; Project administration, A.N.N.; Funding acquisition, A.N.N. All authors have read and agreed to the published version of the mAnuscript.

**Funding:** The Schlumberger Foundation Faculty for the Future Fellowship, and Murdoch University provided financial support for this research, which the authors gratefully acknowledge.

**Data Availability Statement:** The original contributions presented in the study are included in the article, further inquiries can be directed to the corresponding author.

**Conflicts of Interest:** The authors declare no conflict of interest.

## References


1. Phiri, T.C.; Singh, P.; Nikoloski, A.N. The potential for copper slag waste as a resource for a circular economy: A review—Part I. *Miner. Eng.* **2022**, *180*, 107474. [CrossRef]
2. Piatak, N.M.; Parsons, M.B.; Seal, R.R. Characteristics and environmental aspects of slags: A review. *Appl. Geochem.* **2015**, *57*, 236–266. [CrossRef]
3. Phiri, T.C.; Singh, P.; Nikoloski, A.N. The potential for copper slag waste as a resource for a circular economy: A review—Part II. *Miner. Eng.* **2021**, *172*, 107151. [CrossRef]
4. Wang, D.W.; Zhao, Z.W.; Lin, Z.; Liang, Y.J.; Kang, L.; Peng, B. Interaction mechanism between arsenate and fayalite-type copper slag at high temperatures. Trans. *Nonferrous Met. Soc. China* **2022**, *32*, 709–720. [CrossRef]
5. Li, S.; Pan, J.; Zhu, D. A novel process to upgrade the copper slag by direct reduction-magnetic separation with the addition of $Na_2CO_3$ and CaO. *Powder Technol.* **2019**, *347*, 159–169. [CrossRef]
6. Zhang, S.; Zhu, N.; Shen, W.; Wei, X.; Li, F.; mA, W.; mAo, F.; Wu, P. Relationship between mineralogical phase and bound heavy metals in copper smelting slags. *Resour. Conserv. Recycl.* **2022**, *178*, 106098. [CrossRef]
7. Phiri, T.C.; Singh, P.; Yotamu, S.; Hara, Y.S.; Siame, J.; Nikoloski, A.N. Carbothermic Reduction of Copper Slag in the Presence of Borax. *Miner. Process. Extr. Metall. Rev.* **2023**, 1–12. [CrossRef]
8. Yamashita, T.; Hayes, P. Analysis of XPS spectra of $Fe^{2+}$ and $Fe^{3+}$ ions in oxide mAterials. *Appl. Surf. Sci.* **2008**, *254*, 2441–2449. [CrossRef]
9. Li, Y.; Perederity, L.; Papangelakis, V.G. Cleaning of waste smelter slag and recovery of valuable metals by presure oxidative leaching. *J. Hazard. mAter.* **2008**, *152*, 607–615. [CrossRef]
10. Zhang, H.; Hu, C.; Gao, W.; Lu, M. Recovery of Iron from Copper Slag Using Coal-Based Direct Reduction: Reduction Characteristics and Kinetics. *Minerals* **2020**, *10*, 973. [CrossRef]
11. Lan, C.; Zhang, S.; Liu, R.; Lyu, Q.; Yan, G.; Wang, B. Thermodynamic and kinetic behaviours of copper slag carbothermal reduction process. *Ironmak. Steelmak.* **2022**, *50*, 123–133. [CrossRef]
12. Guo, Y.; Zheng, F.; Jiang, T.; Chen, F.; Wang, S.; Qiu, G. Effects of Borax on the Reduction of Pre-oxidized Panzhihua Ilmenite. *JOM* **2017**, *70*, 2018.
13. Gorai, B.; Jana, R.K. Characteristics and utilisation of copper slag—A review. *Resour. Conserv. Recycl.* **2003**, *39*, 299–313. [CrossRef]
14. Vítková, M.; Ettler, V.; Johan, Z.; Kribek, B.; Sebek, O.; Mihaljevic, M. Primary and secondary phases in copper-cobalt smelting slags from the Copperbelt Province. *Zamb. Min. mAg.* **2010**, *74*, 581–600. [CrossRef]
15. Wu, L.; Wang, H.; Dong, K. Effect of Sulfur Content on Copper Recovery in the Reduction Smelting Process. *Metals* **2022**, *12*, 857. [CrossRef]
16. Mackwell, S.J. Oxidation Kinetics of Fayalite ($Fe_2SiO_4$). *Phys. Chem. Miner.* **1992**, *19*, 220–228. [CrossRef]
17. Đorđević, T.; Tasev, G.; Aicher, C.; Potysz, A.; Nagl, P.; Lengauer, C.L.; Pędziwiatr, A.; Serafimovski, T.; Boev, I.; Boev, B. Mineralogy and environmental stability of metallurgical slags from the Euronickel smelter, Vozarci, North Macedonia. *Appl. Geochem.* **2024**, *170*, 106068. [CrossRef]
18. Hu, J.H.; Wang, H.; Zhao, L.M.; Li, L.; Liu, H.L. Characterization of copper slag from impoverishment. *J. Saf. Environ.* **2011**, *11*, 90–93. [CrossRef]
19. Zhao, Z.; Wang, Z.; Peng, N.; Peng, B.; Liang, Y.; Qu, S.; Dong, Z.; Zeng, W. Copper behavior and fayalite microstructure changes influenced by $Cu_2O$ dissolution. *JOM* **2019**, *71*, 2891–2898. [CrossRef]
20. Zhou, S.; Wei, Y.; Li, B.; Wang, H. Effect of Iron Phase Evolution on Copper Separation from Slag Via Coal-Based Reduction. *Metall. mAter. Trans. B* **2018**, *49*, 3086–3096. [CrossRef]
21. Hannon, A.C.; Vaishnav, S.; Alderman, O.L.G.; Bingham, P.A. The structure of sodium silicate glass from neutron diffraction and modeling of oxygen-oxygen correlations. *J. Am. Ceram. Soc.* **2021**, *104*, 6155–6171. [CrossRef]
22. Shi, G.; Liao, Y.; Su, B.; Zhang, Y.; Wang, W.; Xi, J. Kinetics of copper extraction from copper smelting slag by pressure oxidative leaching with sulfuric acid. *Sep. Purif. Technol.* **2020**, *241*, 116699. [CrossRef]
23. Kobylin, P.; Furta, L.; Vilaev, D. *HSC Chemistry 10; Equilibrium Module*; Metso Outotec: Espoo, Finland, 2021.

*metals*

MDPI

*Article*

# Characterisation of Varying Iron Ores and Their Thermal Decomposition Kinetics Under HIsarna Ironmaking Conditions

**Philipp Leerhoff [1,*], Johannes C. Brouwer [1], Amir Mohseni Armaki [1], Christiaan Zeilstra [2], Koen Meijer [2], Jan van der Stel [2], Shoshan T. Abrahami [1], Neslihan Dogan [1] and Yongxiang Yang [1]**

[1] Department of Materials Science and Engineering, Delft University of Technology, 2628 CD Delft, The Netherlands
[2] Tata Steel Netherlands, 1970 CA IJmuiden, The Netherlands
[*] Correspondence: p.leerhoff@tudelft.nl

**Abstract:** In the pre-reduction cyclone of the HIsarna process, both thermal decomposition and gas reduction of the injected iron ores occur simultaneously at gas temperatures of 1723–1773 K. In this study, the kinetics of the thermal decomposition of three iron ores (namely OreA, OreB and OreC) for HIsarna ironmaking were analysed as an isolated process with a symmetrical thermogravimetric analyser (TGA) under an inert atmosphere. Using various methods, the chemical and mineralogical composition, particle size distribution, morphology and phase distribution of the ores were analysed. The ores differ in their mineralogy and morphology, where OreA only contains hematite as iron-bearing phase and OreB and OreC include goethite and hematite. To obtain the kinetic parameters in non-isothermal conditions, the Coats–Redfern Integral Method was applied for heating rates of 1, 2 and 5 K/min and a maximum temperature of 1773 K. The TGA results indicate that goethite and hematite decomposition occur as a two-stage process in an inert atmosphere of Ar. The proposed reaction mechanism for the first stage of goethite decomposition is chemical reaction with an activation energy ranging from 46.55 to 60.38 kJ/mol for OreB and from 69.90 to 134.47 kJ/mol for OreC. The proposed reaction mechanism for the second stage of goethite decomposition is diffusion, showing an activation energy ranging between 24.43 and 44.76 kJ/mol for OreB and between 3.32 and 23.29 kJ/mol for OreC. In terms of hematite decomposition, only the first stage was analysed. The proposed reaction mechanism is chemical reaction control. OreA shows an activation energy of 545.47 to 670.50 kJ/mol, OreB one of 587.68 to 831.54 kJ/mol and OreC one of 424.31 to 592.32 kJ/mol.

**Keywords:** HIsarna ironmaking; iron ore; thermal decomposition; kinetics; extractive metallurgy

## 1. Introduction

The iron- and steelmaking industry alone is responsible for around 7–9% of anthropogenic-generated $CO_2$ emissions [1]. The main reason for this is the blast furnace (BF), which is the dominant operating technology for ironmaking, contributing up to 71% of the global $CO_2$ emissions for steel production [2]. Comparing the currently commercially available established steelmaking routes, BF-BOF, Scrap-EAF, and DRI-EAF, the blast furnace route exhibits the highest amounts of $CO_2$ emissions per tonne of crude steel [2]. Our society is strongly dependent on steel production and considering the increasing production trend over last few decades, steel production is expected to grow further [3], and it is of utmost importance to reduce the $CO_2$ emissions arising from the steel industry.

Iron ores can generally be categorised into high-grade (>65% Fe), medium-grade (62–65% Fe) and low-grade (<62% Fe) ores. The most commercially used iron ores are hematite (69.94% Fe) and magnetite (72.36% Fe), exhibiting the highest amounts of iron content in available ores. Next to hematite and magnetite, multiple other iron ores are present [4]. One of them is goethite, which is abundantly available in India [5]. Goethite ($FeO^*(OH)^*nH_2O$), a medium-grade ore, contains chemically bound water [4]. Alongside the environmental problems arising from the blast furnace route, demand issues in the primary production of steel can become critical. Large amounts of goethite are important to tackle future demand problems in ironmaking industry, but its use leads to technical challenges due to its lower iron content.

In the framework of the ULCOS project, HIsarna (Figure 1) was developed by Tata Steel Netherlands as an alternative ironmaking process with the objective to substitute the blast furnace [6]. HIsarna operates using a counter-current principle, where the injected iron ore is in contact with the hot process gas in a temperature range of 1723–1773 K, thereby combining the two process steps of cyclone converter and smelting reduction in a single operating furnace [7]. In the upper section, the solid particles are heated up, melted and pre-reduced with reducing gas mixtures of $CO$-$CO_2$-$H_2$-$H_2O$, arising from the reactions in the lower section of the furnace [7,8]. In the lower section of the furnace, the final reduction of the pre-reduced molten ores with coal to metallic iron takes place. Due to this reaction, the reducing gas mixture needed in the top section is produced [7]. The degree of pre-reduction that occurs in the top section is around 10–20%. The use of HIsarna, instead of the blast furnace, reduces the three blast furnace process steps, coking, agglomeration and ironmaking, to only one process step, the ironmaking itself. This leads to a lower energy consumption and less $CO_2$ emissions [7]. In combination with carbon capture and storage (CCS) technology, HIsarna is expected to reduce $CO_2$ emissions by up to 80% compared to traditional blast furnace ironmaking [8]. HIsarna has the further advantage of using low-grade, untreated (industrial) ores, with a high amount of gangue material, which are lower in iron content and other residues arising from iron-bearing processes within and outside the steelmaking industry. Further, hydrogen enrichment and the use of renewable carbon sources to improve HIsarna ironmaking technologies are under investigation.

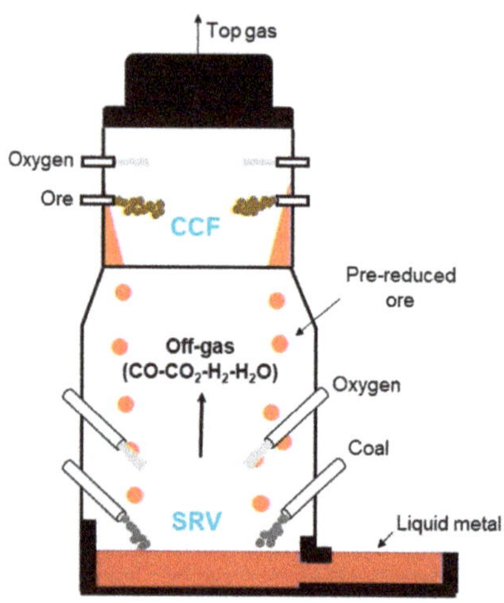


**Figure 1.** Schematic of HIsarna process (Adapted from [7]).

In the top section of a HIsarna furnace, two main reaction mechanisms prevail: thermal decomposition and gas–solid reduction by a gas mixture of $CO$ and $H_2$. During the injection, the fine ores are rapidly heated up, thermally decomposed, pre-reduced by the gas atmosphere and partly melted. Most particles experience short residence times in the top section of the HIsarna furnace, but some particles can hit the walls and become deposited as a molten film. Considering thermodynamics, the reduction of the ores by thermal decomposition and gas reduction has the same effect. Both reactions describe the removal of oxygen from the ore leading to a lower oxygen content. Regarding kinetics, however, both the residence time of the particle in the top section and the mechanisms that occur influence the reaction rates and rate-limiting steps. Qu et al. [9] investigated the thermal decomposition behaviour of hematite ores with low gangue materials and Chen et al. [10] focused on the thermal decomposition behaviour of hematite ores with higher gangue contents. The current study aims to extend this research to the investigation of hematite ores with high contents of gangue materials as well as goethite-based iron ores and to analyse the influence of ore mineralogy on the decomposition kinetics.

## 2. Material Analysis

Varying types of commercial iron-bearing ores were provided by Tata Steel Netherlands, which we called OreA, OreB and OreC. The ores were classified into two different particle size fractions (<63 µm, (63–125) µm) by wet sieving. The particle size is defined as the particle diameter. The size ranges were verified via light scattering technology with the Microtrac Bluewave. Each sample was automatically measured in triplicate. The resulting mean diameter of the volume distribution (MV), number distribution (MN) and area distribution (MA) are presented in Table 1 and the overall particle size distribution in Figure 2. It is important to mention that the three values of MV, MN and MA are used to generate a broad overview on the averaged particle sizes from different perspectives. MV is strongly influenced by coarser particles and therefore usually shows larger average particle sizes. MA is a particle surface measurement and is less affected by coarse particles, leading to smaller average particle sizes. MN is related to the population of the particles and is more sensitive towards small particles, therefore showing the smallest average particle size. Figure 2 portrays the passing percentage in relation to the particle size. A larger passing at low particle sizes, therefore indicates an overall smaller particle size distribution. The <63 µm fractions of OreA, OreB and OreC show almost no particles larger than the defined fraction. The fraction of (63–125) µm OreC shows the largest passing in the defined ranges, followed by OreB and OreA.

**Table 1.** Mean particle size values for varying iron ores.

| | OreA | | OreB | | OreC | |
|---|---|---|---|---|---|---|
| **Fraction [µm]** | <63 | 63–125 | <63 | 63–125 | <63 | 63–125 |
| **MV [µm]** | 50.74 | 103.6 | 44.22 | 100.2 | 22.79 | 115.9 |
| **MN [µm]** | 19.55 | 51.96 | 3.66 | 66.29 | 1.106 | 89.55 |
| **MA [µm]** | 39.41 | 84.86 | 25.52 | 87.58 | 6.80 | 105.5 |

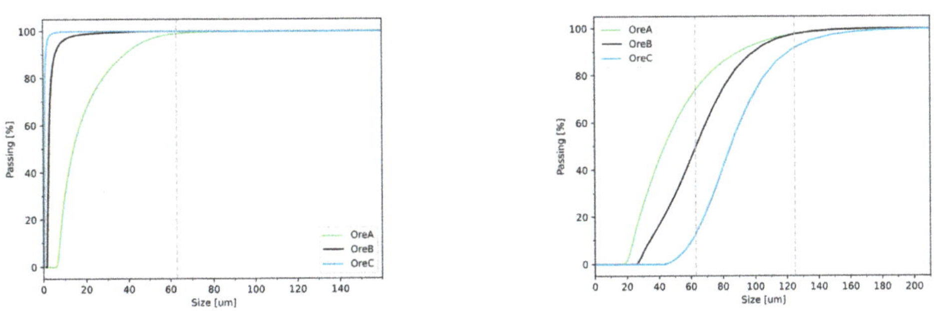


(a) (b)

**Figure 2.** Particle size distribution of OreA, OreB and OreC for size fractions of (**a**) <63 µm and (**b**) (63–125) µm.

### 2.1. Composition Analysis

X-Ray diffraction (XRD) with a Bruker D8 Advance diffractometer Bragg-Brentano geometry and Lynxeye position sensitive detector (Bruker, Germany) was used to evaluate the mineralogy of the samples and to determine the main differences in composition of the different iron ores. In general, the XRD signal is complex, containing a lot of smaller peaks due to the high amount of gangue materials in the ores (Figure S1, Supplementary Materials). For OreA, hematite ($Fe_2O_3$), quartz ($SiO_2$) and dolomite ($CaMg(CO_3)_2$ were detected as the main phases. Additionally, for the fraction < 63 µm a few small magnetite ($Fe_3O_4$) peaks occurred, which were not measured in the other samples. OreB and OreC contained a goethite phase as an iron-bearing mineral in addition to hematite. The mineralogical composition of OreB was goethite (FeOOH), hematite, quartz and kaolinite ($Al_2Si_2O_5(OH)_4$). For OreC only hematite, goethite and quartz were detected as mineral phases.

In order to obtain deeper insights in the elemental composition of the ores, the chemical compositions of the ore fractions were determined by X-Ray fluorescence (XRF) (Panalytical AXIOS XRF, Netherlands) and loss of ignition (LOI) analysis (dissolution in nitric acid and TGA (LECO, Germany)). The results are listed in Table 2, after normalisation and conversion to the expected oxide phases. The total-C in the system was calculated based on the assumption that all measured carbon is present as $CO_2$ in the sample. The content of gangue and iron-bearing phases varies not only between the different ores, but also between the different size fractions. For all ores, the main gangue components were $SiO_2$ and $Al_2O_3$. OreB shows the highest amounts of $Al_2O_3$, underlined by the detected kaolinite phase in the XRD measurements. All ores show similar contents of $SiO_2$, varying in particle size fraction. In contrast to OreB and OreC, the measurements for OreA show higher amounts of CaO and MgO, from the detected dolomite phase. Further, OreB and OreC show lower amounts of carbon compared to OreA, due to the association of carbon to the dolomite phase in OreA. The crystal water in OreB and OreC indicates a goethite phase, with larger amounts present in OreC.

**Table 2.** Chemical composition of varying iron ores based on XRF and LOI.

| | | **OreA** | | **OreB** | | **OreC** | |
|---|---|---|---|---|---|---|---|
| | **Fraction [µm]** | **<63** | **63–125** | **<63** | **63–125** | **<63** | **63–125** |
| **Compound [wt.%]** | $Fe_2O_3$ | 85.93 | 91.50 | 88.07 | 85.07 | 81.21 | 82.92 |
| | total-Fe | 60.10 | 64.00 | 61.60 | 59.50 | 56.80 | 58.00 |
| | $SiO_2$ | 5.18 | 2.96 | 4.96 | 5.83 | 6.84 | 5.30 |
| | $Al_2O_3$ | 2.24 | 1.44 | 3.81 | 4.78 | 3.52 | 3.00 |
| | MgO | 0.71 | 0.41 | 0.04 | 0.04 | 0.14 | 0.11 |
| | CaO | 1.99 | 1.23 | 0.05 | 0.05 | 0.11 | 0.07 |
| | $TiO_2$ | 1.04 | 0.92 | 0.33 | 0.35 | 0.18 | 0.16 |
| | MnO | 1.20 | 0.60 | 0.03 | 0.03 | 1.07 | 1.24 |
| | total-C | 0.41 | 0.23 | 0.06 | 0.07 | 0.18 | 0.14 |
| | $H_2O$(crystal) | - | - | 2.19 | 3.26 | 6.05 | 6.39 |
| | Rest | 0.37 | 0.21 | 0.25 | 0.31 | 0.26 | 0.26 |

### 2.2. Morphological Analysis

To observe the morphology of the different ores, SEM images were taken by a Jeol JSM-IT100 (Jeol, The Netherlands). Representative images of the chosen fractions of the ores are displayed in Figure 3. Significant differences in morphology are visible between the goethite and hematite ores. The morphology of OreA appears to be angular in shape with almost no spherical particles. The structure is close-grained and does not seem to be porous. Bright particles assumed to be further gangue components, are visible either as individual particles or as inclusions on the hematite particles. OreB appears to be less angular and more spherical in structure. The presence of crystal water could be a reason for the apparent porosity of the particles. For OreC, the particles of the larger size fraction (Figure 3f) again appear to be more close-grained, despite the larger amounts of crystal water. Again, the particles shape appears to be more spherical compared to OreA. However, for the <63 μm size fraction (Figure 3c), the particles seem to agglomerate and many fine ore particles are observable as clusters. This is also reflected in the low average particle size values of MV, MN and MA of this fraction. For OreB and OreC no bright particles are visible either on the particle surface or as individual particles.

**Figure 3.** SEM images for (**a**) OreA fraction < 63 μm (150×); (**b**) OreB fraction < 63 μm (150×); and (**c**) OreC fraction < 63 μm (150×); (**d**) OreA fraction (63–125) μm (100×); (**e**) OreB fraction (63–125) μm (100×); and (**f**) OreC fraction (63–125) μm (100×).

### 2.3. *Phase Distribution*

To further investigate the distribution of impurities (gangue) in the samples, the particles of the small fractions (<63 μm) were mounted in a resin, followed by sanding and polishing to create a cross-section of the particles. The cross-sections were again analysed under the SEM, together with EDS mapping to gain insight into the elemental distribution of the particles (Figure 4). Displayed here are only the main elements of the ore, Fe, Al and Si. The mapping of OreA supports the findings from the morphological study, indicating that the distribution of impurities is seen as single, brighter particles, as well as inclusions on the hematite particles. The elemental distribution of OreC indicates a similar distribution compared to OreA. Aluminium and silicon are distributed as distinct particles across the sample, with high intensities of these elements. For OreB, a different elemental distribution is observable. For silicon, a few distinct small particles are visible. Apart from that, the aluminium and silicon compounds seem to be coexisting with the iron ore particles. However, the distribution of iron shows that several iron ore particles are present without the additional presence of aluminium or silicon.

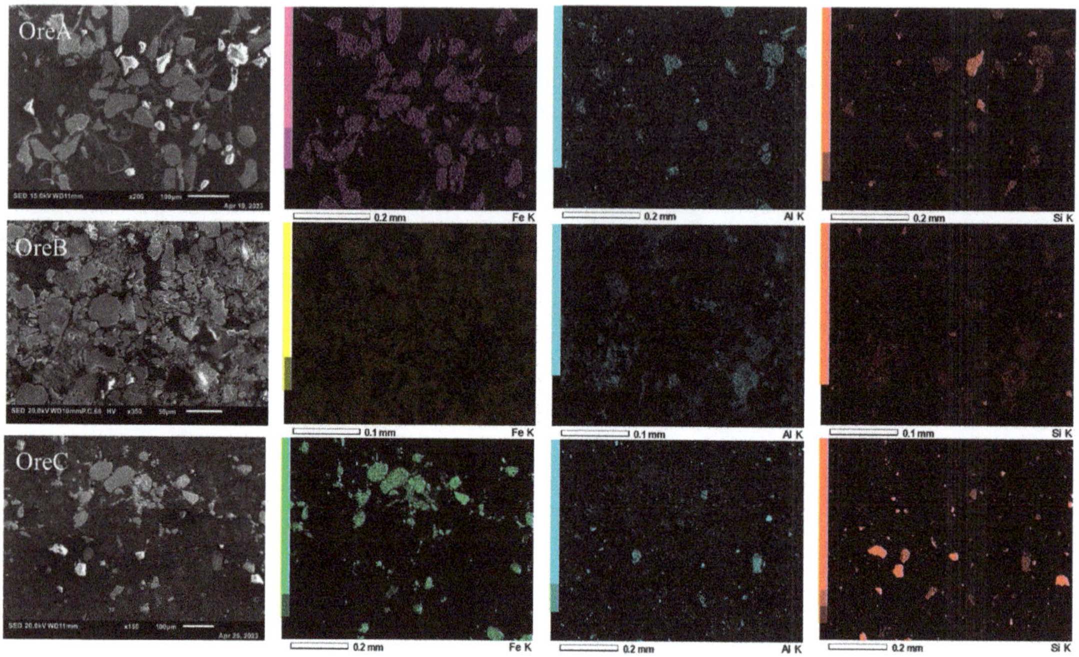


**Figure 4.** Elemental distribution of ore fractions < 63 μm in mounted resin (purple/yellow/green: Fe, blue: Al, red: Si).

It is not possible to distinguish between hematite and goethite using SEM-EDS due to the elemental characteristics of hydrogen. To gain more insights into the distribution of the goethite phase in OreB and OreC, surface-enhanced Raman spectroscopy (alpha300 R-Raman imaging microscope (WITec, Ulm, Germany)) was carried out for both size fractions of OreB and OreC (Figure 5). Data from the literature for pure hematite and goethite [11] were used to identify pure hematite and goethite particles in the samples. Based on the Raman shift in these particles, the reference spectrum presented in Figure 5a was created to map the phase distribution. Next, different representative spots in the samples were measured and compared to the reference spectra. If the shift in the peaks is similar to the reference spectra, it is assumed to be the same phase. Apart from the pure goethite and hematite, a mixed phase was measured, shown in the example spectrum in Figure 5b.

Considering the distribution of goethite and hematite in the large size fraction (63–125 μm), four different particle structures were identified for both OreB and OreC (Figure 5d,f). OreC consists of almost pure hematite and goethite particles, hematite particles with smaller amounts of goethite and goethite particles with distinctive hematite inclusions and a mixed phase. OreB also contains almost pure hematite and goethite particles. However, goethite particles with hematite inclusions and the mixed phase appear to be present in larger quantities. For the smaller size fraction (<63 μm), no individual particles were analysed, but a selected area was scanned for the mapping. It must be noted that the dark area in the scan relates to the resin the particles were embedded in. OreC in Figure 5c shows a distinctive separation of the goethite and hematite phase and small spots of the mixed phase. OreB in Figure 5e also shows a segregation of the hematite and goethite phases. However, the mapping of OreB further indicates the co-existence of goethite and hematite in some areas.

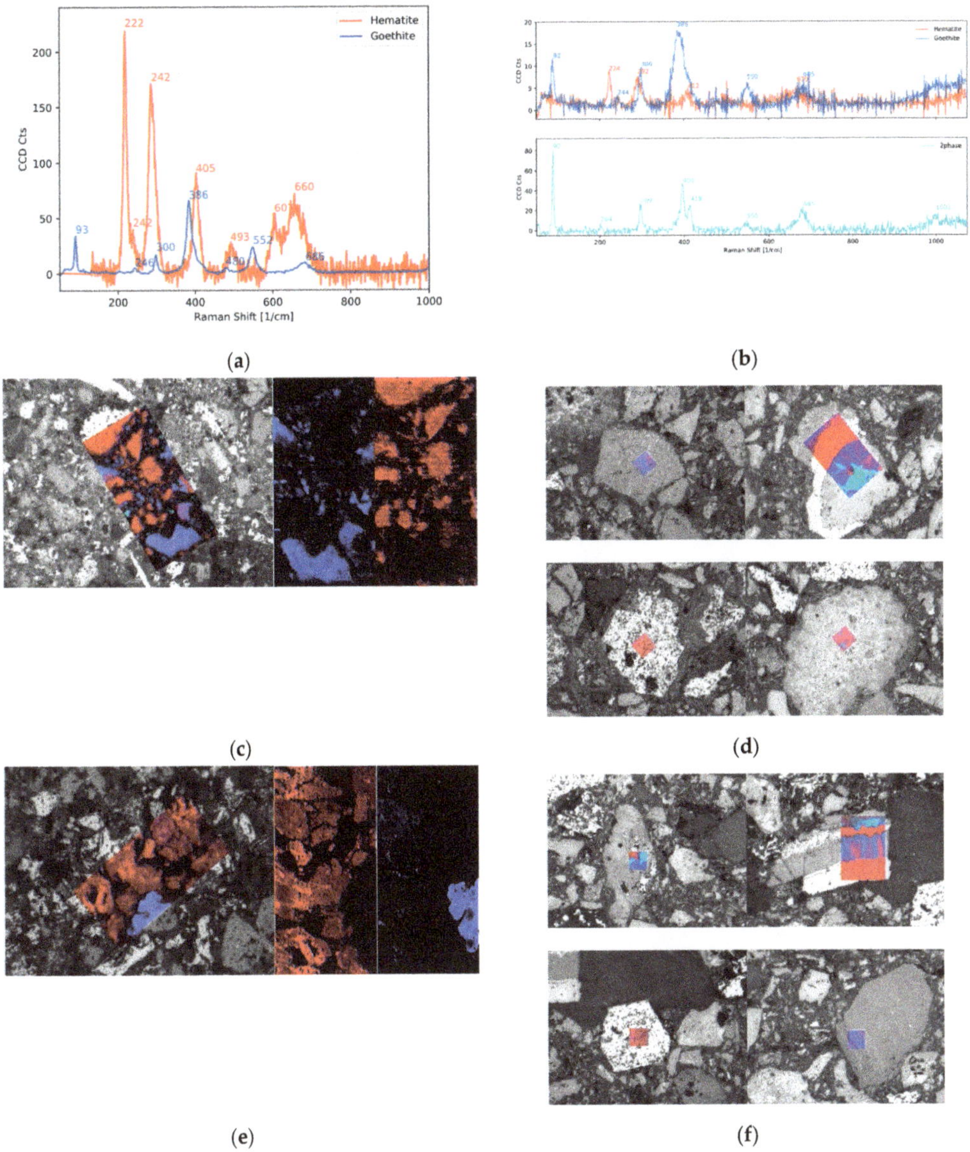


**Figure 5.** Raman spectroscopy for (**a**) the reference spectra; (**b**) the spectra of OreC < 63 μm; (**c**) mapping for OreC < 63 μm (left for complete mapping; right for separated mapping); (**d**) OreC (63–125) μm; (**e**) OreB < 63 μm (left for complete mapping; right for separated mapping); and (**f**) OreB (63–125) μm (red: hematite, dark blue: goethite, light blue: mixed phase).

## 3. Methods

### *3.1. TGA Experiments*

To obtain insights into the non-isothermal decomposition behaviour and kinetics of the goethite and hematite ores, experiments were carried out in a symmetrical thermogravimetric analyser (TGA, Setaram TAG 16/18, Caluire, France). Around 300 mg of ore sample was placed into Alsint crucibles with an inner diameter of 10 mm and a wall thickness of

2 mm. The total height of the crucible is 20 mm. The operating parameters are presented in Table 3. In order to evaluate the thermal decomposition of the ores, the experiments were carried out in an argon atmosphere with a flow rate of 100 mL/min. The ore particles with a size fraction < 63 μm were gradually heated at three heating rates (1, 2 and 5 K/min) until a temperature of 1773 K was reached. Experiments for the particle size fraction of (63–125) μm were only carried out with a heating rate of 2 K/min to compare the results to the smaller particle size fraction at the same heating rate. After 1773 K was reached, a holding time of two hours was applied to achieve the limit of thermal decomposition. After that, the system was cooled to room temperature with a controlled cooling rate of 5 K/min. The mass loss over an increasing temperature at a constant heating rate was measured as the basic data for performing the decomposition and kinetic analysis.

**Table 3.** Experimental conditions in TGA, Setaram TAG 16/18.

| Experimental Condition | Operating Parameter |
|---|---|
| Sample | Hematite fine ore, Goethite fine ore |
| Holding time (h) | 2 |
| Heating rate (K/min) | 1, 2, 5 |
| Particle size (μm) | <63, 63–125 |
| Temperature (K) | 298–1773 |
| Atmosphere | Ar |

### 3.2. *Assessment of Reaction Kinetics of Thermal Decomposition*

The kinetic parameters, activation energy (Ea) and pre-exponential factor A of hematite and goethite decomposition under non-isothermal conditions with a constant heating rate were determined using the Coats–Redfern Integral Method [12]. Through model fitting, the most applicable model describing the thermal decomposition of hematite and goethite was estimated. Commonly used models for gas–solid reactions are listed in Table 4. The essential characteristic parameter for all the kinetic models is the conversion factor α (Equation (1)).

$$\alpha = \frac{m_0 - m_t}{m_0 - m_a} \tag{1}$$

where $m_0$ is the mass of the initial sample (mg), $m_t$ is the mass of the sample at time t (mg) and $m_a$ is the mass of the sample after complete conversion or thermal decomposition (mg). In the following analysis, the final conversion state was considered to be the decomposition from either $Fe_2O_3$ to $Fe_3O_4$ or from FeOOH to $Fe_2O_3$. Hence, a conversion factor of 0 equals the initial state of hematite or goethite and a conversion factor of 1 equals the complete decomposition to magnetite or hematite. The general expression of the reaction rate of the thermal decomposition process is presented in (Equation (2)). Applying the Arrhenius equation, the expression for the reaction rate constant can be determined through (Equation (3)). Considering a constant heating rate, the reaction rate equation can be written as (Equation (4)).

$$\frac{d\alpha}{dt} = kf(\alpha) \tag{2}$$

$$k = A\exp\left(-\frac{E_a}{RT}\right) \tag{3}$$

$$\frac{d\alpha}{dT} = \frac{A}{\beta}\exp\left(-\frac{E_a}{RT}\right)f(\alpha) \tag{4}$$

where k is the reaction rate constant ($s^{-1}$), f(α) is a function of the conversion factor α, A is the exponential factor ($s^{-1}$), $E_a$ is the activation energy (J/mol), β is the heating rate (K/s), T is the temperature (K) and R is the gas constant (8.314 J/(mol·K). The kinetic parameters of hematite and goethite decomposition were determined by the Coats–Redfern Integral Method (Equation (5)).

$$\ln\left[\frac{g(\alpha)}{T^2}\right] = \ln\left[\frac{AR}{\beta E_a}\left(1 - \frac{2RT}{E_a}\right)\right] - \frac{E_a}{RT} \tag{5}$$

where $g(\alpha)$ is the integrated form of the function $f(\alpha)$. The activation energy ($E_a$) and pre-exponential factor (A) can be determined from the slope and intercept of the plot $\ln\left[\frac{g(\alpha)}{T^2}\right]$ vs. $\frac{1}{T}$. Assuming that the value of the activation energy is significantly larger then (2RT), the term $\left(1 - \frac{2RT}{E_a}\right)$ in (Equation (5)) is assumed to equal to 1. Following, the slope of the plot equals $-\frac{E_a}{R}$ and the intercept equals $\ln\left[\frac{AR}{\beta E_a}\right]$. For the calculation of the activation energy, the intercept, reaction rate constant and the resulting reaction rate of the model with the best linear fit ($R^2$ close to 1) was used.

**Table 4.** Reaction models for gas–solid reaction processes (Adapted from [10,13]).

| Reaction Model | | | $f(\alpha)$ | $g(\alpha)$ |
|---|---|---|---|---|
| Nucleation | Power law | P1 | $4\alpha^{3/4}$ | $\alpha^{1/4}$ |
| | | P2 | $3\alpha^{2/3}$ | $\alpha^{1/3}$ |
| | | P3 | $2\alpha^{1/2}$ | $\alpha^{1/2}$ |
| | | P4 | $2/3\alpha^{-1/2}$ | $\alpha^{3/2}$ |
| | Avrami–Erofeev | A2 | $2(1-\alpha)[-\ln(1-\alpha)]^{1/2}$ | $[-\ln(1-\alpha)]^{1/2}$ |
| | | A3 | $3(1-\alpha)[-\ln(1-\alpha)]^{2/3}$ | $[-\ln(1-\alpha)]^{1/3}$ |
| | | A4 | $4(1-\alpha)[-\ln(1-\alpha)]^{3/4}$ | $[-\ln(1-\alpha)]^{1/4}$ |
| Diffusion | n-Dimensional Diffusion | D1 | $1/(2\alpha)$ | $\alpha^2$ |
| | | D2 | $[-\ln(1-\alpha)]^{-1}$ | $(1-\alpha)\ln(1-\alpha)+\alpha$ |
| | | D3 | $3/2(1-\alpha)^{2/3}[1-(1-\alpha)^{1/3}]^{-1}$ | $[1-(1-\alpha)^{1/3}]^2$ |
| Geometrical Contraction | Contracting Area | R2 | $2 \times (1-\alpha)^{1/2}$ | $1-(1-\alpha)^{1/2}$ |
| | Contracting Volume | R3 | $3 \times (1-\alpha)^{2/3}$ | $1-(1-\alpha)^{1/3}$ |
| Reaction order | n-Order | F0 | 1 | a |
| | | F1 | $1-\alpha$ | $-\ln(1-\alpha)$ |
| | | F2 | $(1-\alpha)^2$ | $(1-\alpha)^{-1}-1$ |
| | | F3 | $(1-\alpha)^3$ | $[(1-\alpha)^{-2}-1]/2$ |

## 4. Results and Discussion

Even though most fine ores injected into the HIsarna furnace experience very short residence times (ms) in the pre-reduction cyclone of the system, particles can agglomerate in clusters or deposit on the walls of HIsarna furnace, leading to longer particle residence times. Gas reduction and thermal decomposition are expected to occur as co-existing phenomena. This paper studies the limits of thermal decomposition as an isolated mechanism of the presented ores. The Results section will first focus on the general thermal decomposition behaviour of the varying ores to identify the different decomposition stages and their temperature ranges. For this, the mass loss and QMS (quadrupole mass spectrometer) off-gas data will be presented. Based on the mass loss's dependency on temperature, a kinetic analysis is further carried out. First, the conversion factor will be determined as an underlying kinetic parameter for the defined temperature ranges and applied to the model fit. Based on the model fit, the activation energy for the varying iron ores will be compared.

### 4.1. Thermal Decomposition

Figure 6 shows different mass loss curves from the thermogravimetric measurement of the feed materials with particle size fractions of OreA, OreB and OreC with varying heating rates (Figure 6a) and a comparison to the larger particle size of (63–125) µm is shown in Figure 6b. Three mass loss stages are observed for OreA, while four stages are seen for OreB and OreC. The initiation of the individual mass loss curves shows slight dependencies on the heating rate for all ores. The mass loss and therefore the decomposition starts earlier with lower heating rate. Larger dependencies are visible for the ores based on mineralogy. OreC shows the highest mass loss, while OreA shows the lowest mass loss under the same thermal conditions.

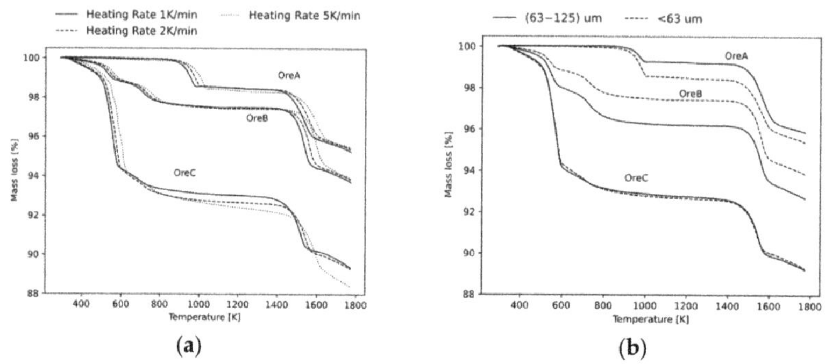


**Figure 6.** Mass loss curves of hematite and goethite ore particles: (**a**) fraction of <63 μm for varying heating rates and (**b**) fraction of <63 μm and (63–125) μm for a heating rate of 2 K/min in non-isothermal conditions.

To obtain details regarding the decomposing gas phases, the off-gas was measured via QMS. Examples of the mass loss curves in combination with the QMS data are presented in Figure 7 for each ore at a heating rate of 2 K/min. Only the main measured gas phases are presented. It must be mentioned that the presented analysis only provides qualitative data, meaning the height of the peaks cannot be compared between experiments. All ores show a distinct $O_2$ peak at later stages in the reaction. This peak indicates the decomposition of the hematite phase under the removal of oxygen. Above temperatures of 570 °C, the reduction of hematite ($Fe_2O_3$) occurs step-wise, initially forming porous magnetite ($Fe_3O_4$) and later wüstite ($Fe_{(1-x)}O$) [14]. Additionally, OreA contains a distinct peak of $CO_2$ emission. Based on the initial composition of OreA, this peak reflects the decomposition of carbonates (dolomite). The decomposition of dolomite occurs at low $CO_2$ partial pressures as a single step reaction (Equation (6)) [15].

$$CaMg(CO_3)_2 \rightarrow CaO + MgO + 2CO_2 \quad (6)$$

This behaviour of OreA was also observed in the publication of Chen et al. [10], in which the first mass loss stage was described as calcination (thermal decomposition) of carbonates and the two later mass loss stages were described as the two-stage decomposition of hematite. The QMS signals of OreB and OreC further indicate the emission of water vapour. Water vapour is emitted as the goethite phase converts into hematite based on (Equation (7)) [16].

$$2\alpha FeOOH \rightarrow \alpha Fe_2O_3 + H_2O \quad (7)$$

OreC shows a generally higher mass loss due to the larger amount of goethite in the raw material, leading to a larger (yet, sharp) emission of water vapour. A more continuous emission of water vapour is observed for OreB, indicating a stronger inclusion of the compound in the ore matrix. This behaviour is supported by the Raman spectroscopy presented in Figure 5. It is expected that the goethite present as individual particles is more accessible and therefore decomposes earlier. Consequently, the first stronger peak, which is present for both ores, is assigned to the decomposition of the almost pure goethite particles in the sample. The second peak, as seen in Figure 7b and to lesser extend in Figure 7c, is presumed to be the decomposition of the goethite from the mixed phase or the goethite which is intertwined with hematite in the particles.

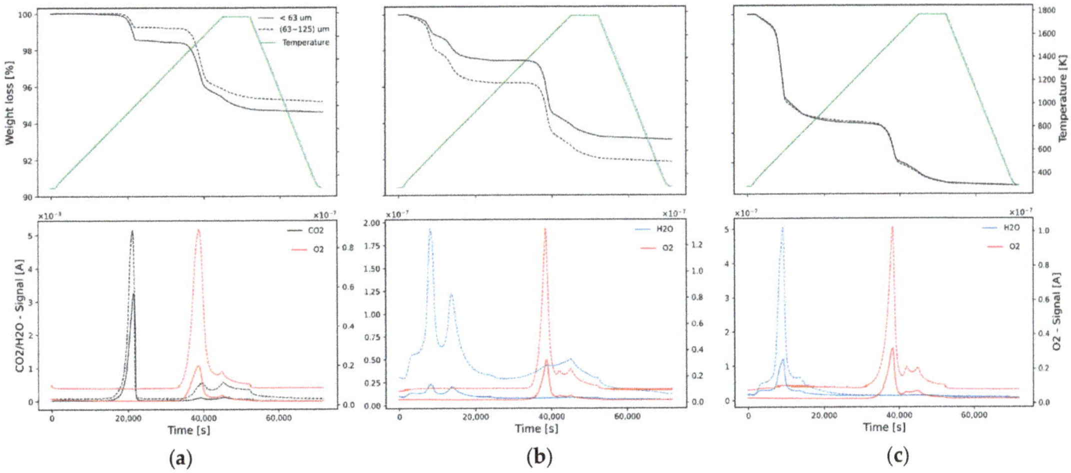

**Figure 7.** TGA data in combination with temperature profile and QMS data for (**a**) OreA, (**b**) OreB and (**c**) OreC for a heating rate of 2 K/min (solid line: <63 µm size fraction, dotted line: (63–125) µm size fraction).

The start and end temperatures for the individual decomposition stages are determined by the temperature boundaries of the first derivative of the mass loss curves, presented in Table 5 in combination with the measured mass loss in the respective stages. Walter et al. [17] investigated conversion temperatures of goethite to hematite for different particle sizes and observed conversion temperatures between 465 and 523 K. Depending on the particle volume and surface area, they further found either one or two peaks arising in the differential thermal analysis (DTA) measurements. The maximum of the first peak varied between 545 and 551 K and that of the second peak between 578 and 595 K. It is observed in here that, for OreB and OreC, the temperature of the two goethite decomposition stages varies between the two ores. For OreB the decomposition appears in two steps with a similar wide temperature range. Depending on the heating rate, the first stage initiates between 385 and 436 K and ends between 595 and 631 K, exhibiting a mass loss of 0.56% to 0.67%. In comparison, the first decomposition stage of OreC initiates between 395 and 502 K and ends between 616 and 680 K, with a respective mass loss of 3.81% to 4.10%. For OreC, a comparably low mass loss of 0.13 to 0.20% is observed in the second stage. For OreB, the mass loss is similar compared to the first stage (0.55 to 0.65%), with initiation temperatures of 600 to 634 K and ending temperatures of 834 to 917 K. The general higher mass loss of OreC is explained by the higher goethite content in the original ore. Further, the ores vary in average particle size and morphology, possibly influencing the initiation of the peaks.

Chen et al. [10] defined two temperature regions for hematite decomposition. The first stage ranges from 1423 to 1593 K and the second stage is defined as being above 1593 K. Qu et al. [9] observed a similar temperature range for hematite decomposition between 1473 and 1573 K. This study observed slightly lower initiation temperatures for hematite decomposition, between 1312 and 1419 K. Contrary to this study, Chen et al. and Qu et al. used lower amounts of input material and samples differing in particle size and mineralogy from this study, which could be reasons for the variation in the decomposition temperature. Further, Qu et al. only tested a heating rate of 10 K/min and Chen et al., heating rates between 2 and 10 K/min, whereas a higher heating rate usually moves the onset of decomposition to higher temperatures. In the first stage of hematite decomposition, the corresponding mass loss is between 1.95% and 2.66%. The second stage, which begins at temperatures between 1676 and 1743 K, records a significantly lower mass loss, ranging between 0.06% and 0.38%, signifying that most thermal decomposition occurs in the first stage. Chen et al. [10] compared the significantly larger first stage to the pre-reduction of a particle in suspension, forming clusters with a molten slag phase. The second stage was described to be the reaction of the ores accumulating at the walls of HIsarna, which is in alignment with the differences in weight loss of the two hematite decomposition stages in this study.

**Table 5.** Decomposition stages of OreA, OreB and OreC for the particle size fraction < 63 μm.

| | | OreA Heating Rate [K/min] | | | | | OreB Heating Rate [K/min] | | | OreC Heating Rate [K/min] | | |
|---|---|---|---|---|---|---|---|---|---|---|---|---|
| | | **1** | **2** | **5** | | | **1** | **2** | **5** | **1** | **2** | **5** |
| **Stage I (Carbonate decomposition)** | Start_T [K] | 743 | 719 | 796 | **Stage I (Goethite decomposition)** | Start_T [K] | 385 | 420 | 436 | 395 | 429 | 502 |
| | End_T [K] | 1053 | 1047 | 1125 | | End_T [K] | 595 | 616 | 631 | 616 | 647 | 680 |
| | Mass loss [%] | 1.38 | 1.38 | 1.43 | | Mass loss [%] | 0.67 | 0.56 | 0.65 | 4.10 | 4.06 | 3.81 |
| **Stage II (Hematite decomposition)** | Start_T [K] | 1325 | 1397 | 1382 | **Stage II (Goethite decomposition)** | Start_T [K] | 600 | 623 | 634 | 621 | 670 | 687 |
| | End_T [K] | 1687 | 1625 | 1724 | | End_T [K] | 834 | 871 | 918 | 793 | 784 | 830 |
| | Mass loss [%] | 2.28 | 1.95 | 2.24 | | Mass loss [%] | 0.55 | 0.62 | 0.65 | 0.20 | 0.13 | 0.14 |
| **Stage III (Hematite decomposition)** | Start_T [K] | 1699 | 1708 | 1726 | **Stage III (Hematite decomposition)** | Start_T [K] | 1331 | 1353 | 1419 | 1312 | 1375 | 1376 |
| | End_T [K] | - | - | - | | End_T [K] | 1625 | 1647 | 1671 | 1588 | 1623 | 1695 |
| | Mass loss [%] | 0.26 | 0.23 | 0.06 | | Mass loss [%] | 2.66 | 2.55 | 2.23 | 2.37 | 2.26 | 2.09 |
| | | | | | **Stage IV (Hematite decomposition)** | Start_T [K] | 1721 | 1720 | 1742 | 1676 | 1708 | 1705 |
| | | | | | | End_T [K] | - | - | - | - | - | - |
| | | | | | | Mass loss [%] | 0.18 | 0.13 | 0.05 | 0.38 | 0.14 | 0.06 |

## *4.2. Kinetics of Goethite Decomposition*

### 4.2.1. Influence of Ore Mineralogy

To analyse the conversion of goethite to hematite through the decomposition of OreB and OreC, the conversion results for the particle size fraction of <63 μm, are plotted in Figure 8a. Figure 8b shows the first derivative of the weight loss curves explaining the conversion stages. Two distinctive peaks are visible for both ores, indicating the two decomposition stages of the goethite phase. A conversion factor of 1 equals complete decomposition of the goethite phase to hematite under the release of water vapour, as presented in Equation (7). The initial decomposition of the goethite phase is comparably slow but increases steeply at around 500 K. Due to the strong decomposition of OreC in the first stage, the conversion is steeper throughout the first stage and almost reaches full conversion in this stage. This increase occurs in a short temperature range, and a conversion factor above 0.8 is reached. The conversion in the second stage continues more slowly, until full conversion is reached. Compared to that, OreB conversion occurs over a larger temperature range with a significant amount of conversion occurring in the second stage of goethite thermal decomposition. The differences in conversion can again be explained by the results of the Raman spectroscopy in Figure 5. First, a steep increase in conversion can be expected due to the initial decomposition of almost pure and, therefore, more accessible goethite particles. Since the goethite content in OreC is significantly higher compared to OreB, more conversion happens at lower temperatures, while the conversion of the mixed phases occurs at a later stage.

It is important to understand how the decomposition behaviour influences the decomposition kinetics. The most essential kinetic parameter to determine is the reaction rate constant, containing the activation energy and the pre-exponential factor. The model fitting approach, explained in the previous section, has been applied to the different experimental results regarding goethite decomposition. Based on the best linear fit, the rate-determining models can be evaluated. The $R^2$ for the linear fit of the different decomposition experiments are presented in Table S1 in the supplementary materials. The bold values of the fitting mark the reaction model, which has been used for further calculations. A number of models were not able to present the experimental data in the second stage of goethite decomposition, due to them violating the assumption that the activation energy needs to be larger than 0. The assumption-violating models for OreB are the nucleation models P1, P2 and P3 and for OreC they are the nucleation models P1, P2, P3 and P4 and the zero-order reaction model F0. For both ores, the first stage of goethite decomposition appears to be limited by the reaction order, followed by diffusion control for the second stage of goethite decomposition.

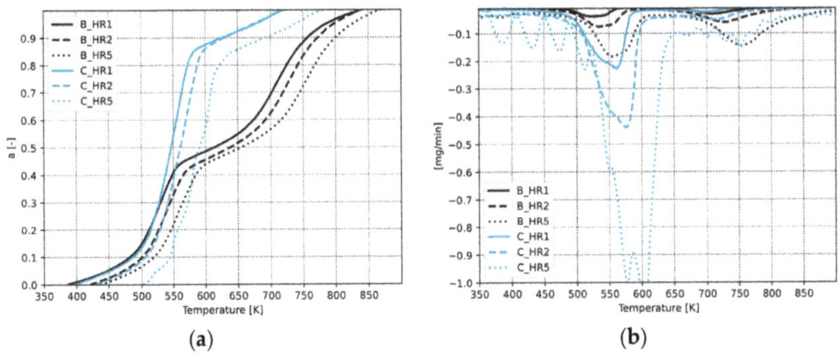


**Figure 8.** Dependency of temperature on (**a**) conversion factor of goethite decomposition for OreB and OreC at varying heating rates (HR) and (**b**) the first derivate of the representative weight loss curves.

Based on the chosen models, the activation energy (Ea) and pre-exponential factor (A) were calculated in order to describe the reaction rate constant (Table 6). The reaction rate constant shows differences depending on the heating rate and ore mineralogy. For OreB the activation energy increases in stage I with the increasing heating rate (46.55 kJ/mol, 56.44 kJ/mol, 60.38 kJ/mol) and decreases with the increasing heating rate in stage II (44.76 kJ/mol, 33.38 kJ/mol, 32.36 kJ/mol). In the case of OreC, no clear trend is observed for the activation energy in stage I. The heating rate of 5 K/min still shows the highest activation energy (134.47 kJ/mol), but the activation energy for the heating rate of 2 K/min (77.86 kJ/mol) is lower compared to the heating rate of 1 K/min (102.62 kJ/mol). The reason for this might be the model selection in the experiment with a heating rate of 1 K/min. For the second decomposition stage, the activation energy again decreases with an increasing heating rate (23.29 kJ/mol, 8.60 kJ/mol, 7.79 kJ/mol). The influence of the heating rate on the activation energy is also observed in further studies [10,18,19]. A difference in the activation energy is visible between the two ores. OreC shows a much higher activation energy in the first decomposition stage and OreB in the second, implying that the two-stage decomposition of OreB is more significant.

**Table 6.** Reaction rate constant for two stage goethite decomposition of OreB and OreC with respect to heating rate.

| **Experiments** | | **Stage I** | | **Stage II** | |
|---|---|---|---|---|---|
| **Ore** | **Heating Rate [K/min]** | **Model** | **Rate Constant** | **Model** | **Rate Constant** |
| OreB | 1 | F3 | $k = 369.27\frac{1}{min}\times\exp(-\frac{46{,}550\frac{J}{mol}}{RT})$ | D3 | $k = 3.06\frac{1}{min}\times\exp(-\frac{44{,}759\frac{J}{mol}}{RT})$ |
| OreB | 2 | F3 | $k = 5002.42\frac{1}{min}\times\exp(-\frac{56{,}441\frac{J}{mol}}{RT})$ | D2 | $k = 1.48\frac{1}{min}\times\exp(-\frac{33{,}378\frac{J}{mol}}{RT})$ |
| OreB | 5 | F3 | $k = 19.71\times10^{3}\frac{1}{min}\times\exp(-\frac{60{,}382\frac{J}{mol}}{RT})$ | D2 | $k = 2.40\frac{1}{min}\times\exp(-\frac{32{,}357\frac{J}{mol}}{RT})$ |
| OreC | 1 | D3 | $k = 9.25\times10^{6}\frac{1}{min}\times\exp(-\frac{102{,}620\frac{J}{mol}}{RT})$ | D3 | $k = 0.17\frac{1}{min}\times\exp(-\frac{23{,}289\frac{J}{mol}}{RT})$ |
| OreC | 2 | F2 | $k = 1.24\times10^{6}\frac{1}{min}\times\exp(-\frac{77{,}860\frac{J}{mol}}{RT})$ | D2 | $k = 0.016\frac{1}{min}\times\exp(-\frac{8459\frac{J}{mol}}{RT})$ |
| OreC | 5 | F3 | $k = 2.96\times10^{11}\frac{1}{min}\times\exp(-\frac{134{,}465\frac{J}{mol}}{RT})$ | D2 | $k = 0.025\frac{1}{min}\times\exp(-\frac{7787\frac{J}{mol}}{RT})$ |

Studying the non-isothermal decomposition kinetics of commercial goethite ores, Walter et al. [17] observed activation energies varying between 107.4 and 137.8 kJ/mol for an overall dehydration of goethite, which they compared to the activation energy found by Pelino et al. [20] (119 kJ/mol) who investigated the decomposition of α-goethite. The activation energies found for the first decomposition stage of OreC are in a similar range, indicating that a difference in ore mineralogy does not have a significant influence on the activation energy. However, considering OreB, with lower goethite contents, the activation energy is lower for both individual stages compared to the overall activation energy values found in the literature. Beuria et al. [21] investigated the isothermal decomposition of iron ores with a goethite content of 47.41 wt.% and found an activation energy ranging between 58.74 and 65.75 kJ/mol, slightly higher compared to stage I of OreB.

4.2.2. Influence of Particle Size

In order to compare the influence of particle size on the decomposition kinetics of industrial goethite ores, TGA experiments of the particle size fractions <63 μm and (63–125) μm were carried out at a heating rate of 2 K/min. Figure 9 compares the conversion factors of OreB and OreC for the two particle size fractions at a heating rate of 2 K/min. The results show that the conversion rate depends more on the goethite content than on the particle size. As concluded from the XRD and XRF measurements in Section 2.1, the amount of crystal water in OreC is similar for both particle sizes (approx. 6 wt.%). In accordance, the course of the conversion over temperature is similar. The results of the particle size comparison of OreB indicate that higher amounts of crystal water can lead to earlier conversion of the goethite phase. The amount of crystal water in the sample of the larger fraction is 3.26 wt.%, compared to 2.19 wt.% in the smaller size fraction. This results in an earlier conversion of the (63–125) μm size fraction. With the increase in particle size, the initiation temperature of the first stage of goethite decomposition of OreB decreases from 420 to 407 K and for the second stage from 623 to 616 K.

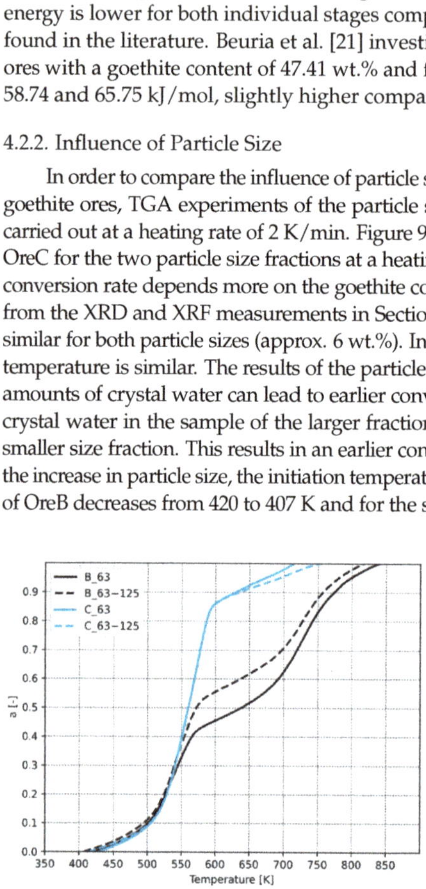


**Figure 9.** Dependency of particle size on the conversion factor at a heating rate of 2 K/min.

The model fitting approach leads to the same rate-determining mechanism for the different particle size fractions of the same ore (Stage I: chemical reaction; Stage II: diffusion). The kinetic data is displayed in Table 7. For OreC, there is a negligible effect of particle size on the conversion factor. As discussed in Section 2.3, the goethite phase is present as a mixed phase with hematite and as a pure phase. The identical reaction mechanisms between the two size fractions indicate that the reaction mechanism is independent of the particle size and more dependent on the nature of the ore. Comparing the reaction rate constants of the larger particle size fractions, a slight influence of the particle size on the activation energy of the goethite decomposition in OreC becomes visible. The activation energy for the larger size fraction (63–125) μm size fraction of OreC is lower for stage I (69.90 kJ/mol) and stage II (3.32 kJ/mol) compared to the activation energy of the smaller size fraction < 63 μm at a heating rate of 2 K/min. For OreB, the activation energy of the (63–125) μm size fraction in stage I (56.26 kJ/mol) is similar compared to the smaller fraction. However, the activation energy calculated for stage II of the larger size fraction of

(63–125) µm, 24.43 kJ/mol, is significantly lower compared to the smaller size fraction. The lower activation energy indicates that the goethite present as a mixed phase with hematite is easily accessible for the larger particle size fraction. The similar activation energy in the first stage of goethite decomposition of OreB indicate a similar behaviour of the initially decomposing pure goethite particles in the samples.

**Table 7.** Reaction rate constants for two-stage goethite decomposition of OreB and OreC with a size fraction of (63–125) µm.

| **Stage I** | | | | |
|---|---|---|---|---|
| Ore | Heating Rate [K/min] | $R^2$ | Model | Rate Constant |
| OreB | 2 | 0.9653 | F3 | $k = 6659.33 \frac{1}{min} \times exp(-\frac{56{,}254 \frac{J}{mol}}{RT})$ |
| OreC | 2 | 0.9657 | F2 | $k = 207273.76 \frac{1}{min} \times exp(-\frac{69{,}901 \frac{J}{mol}}{RT})$ |
| **Stage II** | | | | |
| Ore | Heating rate [K/min] | $R^2$ | Model | Rate constant |
| OreB | 2 | 0.9461 | D2 | $k = 0.32 \frac{1}{min} \times exp(-\frac{24{,}431 \frac{J}{mol}}{RT})$ |
| OreC | 2 | 0.9043 | D2 | $k = 0.0024 \frac{1}{min} \times exp(-\frac{3316 \frac{J}{mol}}{RT})$ |

### 4.3. Kinetics of Hematite Decomposition

The start temperatures in Table 5 (stage II for OreA and stage III for OreB and OreC) define the beginning of the hematite thermal decomposition. The lowest measured initiation temperature is 1312 K. The non-isothermal decomposition zone is the temperature range of the start temperature in the respective stage until the set temperature of 1773 K is reached. The isothermal zone is the following fixed temperature of 1773 K at a holding time of 2 h. In order to first evaluate the differences in the decomposition behaviour between the ores, the total degree of reduction R[%] at the end of the non-isothermal zone and the total degree of reduction at the end of the isothermal period are presented in Figure 10. The degree of reduction describes how much of the hematite is converted to metallic iron or its intermediate products of magnetite and wüstite, due to the removal of oxygen. The figure shows the degree of reduction depending on time and compares the three ores at individual heating rates. A degree of reduction of 11.11% would indicate a complete conversion from hematite to magnetite. A further increase in the degree of reduction to 33.33% would consequently indicate a full conversion to wüstite. Depending on the heating rate, the degree of reduction of the <63 µm fraction of OreA varies between 9.48 and 10.75% at the end of the non-isothermal zone and between 11.82 and 12.42% at the end of the isothermal zone. These reduction degrees are similar to the reduction degrees achieved by Qu et al. [9] for hematite ores (11.8%). The degrees of reduction at the beginning of the isothermal zone are higher for OreB (11.90–12.90%) and OreC (11.91–13.81%) and increase further until the end of the isotherm (OreB: 14.23–14.57%, OreC: 14.18–16.49%). Comparing the degrees of reduction of the <63 µm size fraction to the (63–125) µm size fraction at a heating rate of 2 K/min, no significant increase in the total degree of reduction is observed for any of the ores (OreA: 12.33% → 12.82%, OreB: 14.23% → 14.34%, OreC: 14.18% → 14.35%). For all ores, the reduction by thermal decomposition is almost completed at the beginning of the isothermal period. The reduction degrees of all the ores indicate that a complete conversion to magnetite occurred and small amounts of wüstite should be present.

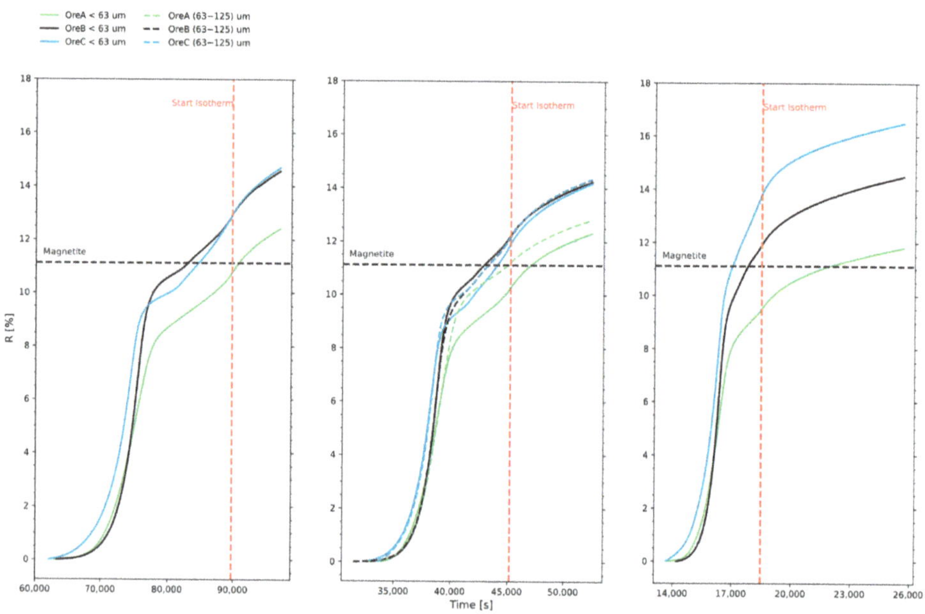


(a) (b) (c)

**Figure 10.** The degree of reduction at the end of the non-isothermal and isothermal heating zone for varying ores and particle sizes at a heating rate of (**a**) 1 K/min, (**b**) 2 K/min and (**c**) 5 K/min (see Table 5 for starting temperatures of hematite decomposition).

In order to validate the presence of the magnetite phase for the further kinetic calculations, example XRD measurements of decomposition products of the size fraction < 63 µm were carried out. For this, a cross-section sample of the solidified material in the crucibles was prepared for analysis. The samples were cut with the Accutom-100 by Struers. Since it was not possible to remove the solidified samples from the crucible, the cross-section was made through the whole crucible. The Alsint crucibles are brittle in nature. Therefore, the cutting spot varies between the samples. The material was solidified by cooling the material in argon atmosphere with a cooling rate of 5 K/min back to room temperature. SEM images of the cross-sections are presented in Figure 11. The sample in the crucible presented in Figure 11a,c was cut vertically, whereas the sample in Figure 11b was cut horizontally.

(a) (b) (c)

**Figure 11.** SEM images of cross-sections of (**a**) OreA, <63 µm, heating rate of 2 K/min, (**b**) OreB, <63 µm, heating rate of 2 K/min and (**c**) OreC, <63 µm, heating rate of 5 K/min.

For all observed ores, the cross-section appears similar, showing solidified pre-reduced particles. In contrast to the original ores (Figure 3), the particles now appear in a similar angular shape. The similar shape indicates that the hematite and goethite transitioned to the same iron oxide phase, regardless of the varied initial ore phases. The XRD pattern are displayed in Figure S2 in the supplementary materials. For the samples in Figure 11a,b, the XRD measurements of these cross-sections confirm magnetite as the only iron oxide phase. For the sample in Figure 11c small amounts of fayalite ($Fe_2SiO_4$) were detected in the XRD pattern next to the magnetite phase. Following the XRD results, magnetite was chosen as the final product for the calculation of the conversion factor of hematite decomposition (Figure 12).

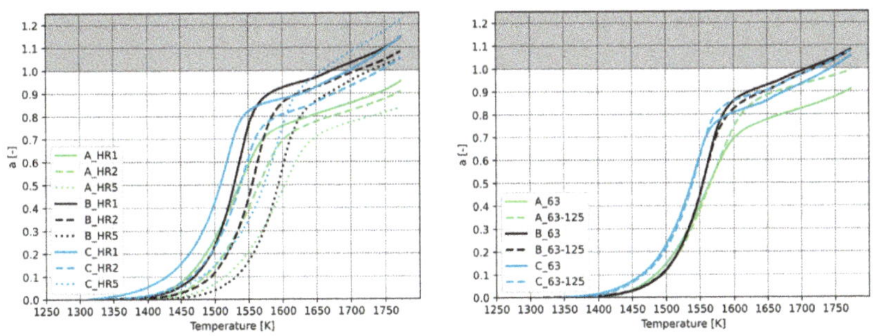


(**a**) (**b**)

**Figure 12.** Dependency of temperature on conversion factor of hematite decomposition for (**a**) varying heating rates (HR) for a particle size of <63 μm and (**b**) for a heating rate of 2 K/min for particle size of (63–125) μm and <63 μm.

In the following section, only the non-isothermal decomposition kinetic parameters will be calculated. Therefore, for the calculations of the conversion factor the isothermal zone is disregarded. Due to this, the smaller size fraction of OreA does not reach full conversion to magnetite in the presented temperature range. The larger particle size, however, leads to a faster conversion and has reached full conversion to magnetite in the non-isothermal heating zone. Due to the higher reduction degrees in the non-isothermal zone, full conversion to magnetite is reached before the beginning of the isothermal zone for OreB and OreC. The grey area of the figure marks when the conversion factor exceeds unity and the conversion to magnetite is completed and the conversion to wüstite has started already. This area is excluded from the kinetic analysis. OreC shows a faster conversion compared to OreB at all heating rates. In general, a lower heating rate leads to an earlier conversion to magnetite. Model fitting was again applied in order to analyse the present reaction mechanisms of hematite decomposition. As shown in Table 5, most of the decomposition occurs during the first stage (third peak of first derivative) and the influence of the second stage (fourth peak of first derivative) is comparably negligible. Due to this, only the first stage of hematite decomposition is analysed in terms of kinetics. Based on the best linear fit, the rate determining models are evaluated. The $R^2$ values for the linear fit for the different decomposition experiments for the particle size fraction < 63 μm are presented in Table S2 in the supplementary materials. The bold values in the fitting mark the reaction model, which has been used for further calculations.

Independent of ore mineralogy and heating rate, all model fittings indicate chemical reaction control as reaction mechanism. The hematite decomposition of OreA is a third-order reaction for all heating rates. For OreB and OreC the heating rate influences the order of reaction. For both ores, a heating rate of 1 or 2 K/min leads to a second-order reaction. For OreB, a heating rate of 5 K/min leads to a third-order and for OreC to a first-order reaction. These results contradict the findings of Chen et al. [10], who described the reaction mechanism of hematite decomposition in a temperature range of 1425–1593 K in non-isothermal conditions as the inner diffusion of iron in the solid phase. The results of our study are more in alignment with the findings of Xing et al. [18], who also found that a second-order reaction as most applicable to the experimental data. The reason for this may be the use of the applied method. As in this study, Xing et al. [18] applied the Coat–Redfern method to determine the kinetic parameters, whereas Chen et al. [10] chose iso-conversional methods to obtain kinetic parameters, before applying them to determine the reaction mechanism. Based on the best-fitting models, the reaction rate constants of the present study were determined and are presented in Table 8.

**Table 8.** Reaction rate constant of the first stage of hematite decomposition for particle size of <63 μm.

| Ore | Heating Rate [K/min] | Model | Rate Constant |
|---|---|---|---|
| A | 1 | F3 | $k = 1.01 \times 10^{17} \frac{1}{min} \times exp(-\frac{545{,}473 \frac{J}{mol}}{RT})$ |
| A | 2 | F3 | $k = 2.01 \times 10^{21} \frac{1}{min} \times exp(-\frac{667{,}451 \frac{J}{mol}}{RT})$ |
| A | 5 | F3 | $k = 7.8 \times 10^{17} \frac{1}{min} \times exp(-\frac{573{,}024 \frac{J}{mol}}{RT})$ |
| B | 1 | F2 | $k = 9.44 \times 10^{20} \frac{1}{min} \times exp(-\frac{654{,}981 \frac{J}{mol}}{RT})$ |
| B | 2 | F2 | $k = 7.27 \times 10^{20} \frac{1}{min} \times exp(-\frac{655{,}272 \frac{J}{mol}}{RT})$ |
| B | 5 | F3 | $k = 7.77 \times 10^{26} \frac{1}{min} \times exp(-\frac{831{,}543 \frac{J}{mol}}{RT})$ |
| C | 1 | F2 | $k = 5.47 \times 10^{15} \frac{1}{min} \times exp(-\frac{498{,}181 \frac{J}{mol}}{RT})$ |
| C | 2 | F2 | $k = 7.13 \times 10^{15} \frac{1}{min} \times exp(-\frac{505{,}468 \frac{J}{mol}}{RT})$ |
| C | 5 | F1 | $k = 1.03 \times 10^{13} \frac{1}{min} \times exp(-\frac{424{,}311 \frac{J}{mol}}{RT})$ |

Comparing the heating rates of the various ores, no distinct effect of the heating rate is observable. A larger influence is seen for the ore mineralogy. The lowest activation energies are present for OreC, ranging between 424.31 kJ/mol and 505.47 kJ/mol. Higher activation energies were observed for OreA (545.47 kJ/mol–667.45 kJ/mol) and for OreB (654.98 kJ/mol–831.54 kJ/mol). Despite the different methods applied, the activation energy of the first stage of hematite decomposition is in alignment with the activation energy determined by Chen et al. [10] for the first stage of hematite decomposition (636 kJ/mol). The activation energy in the study of Xing et al. [18] was much higher with an average of 1256 kJ/mol. The reason for this may be that the model fit was applied to the overall reaction and not the individual stages of hematite decomposition.

Table 9 presents the reaction rate constants for the larger particle size fractions. The reaction mechanism did not change with the increase in particle size. Directly comparing the results to the smaller size fractions, OreA and OreC show a higher activation energy at an increased particle size. Interestingly, the activation energy of OreB decreases with an increase in particle size to 587.68 kJ/mol at a heating rate of 2 K/min.

**Table 9.** Reaction rate constant in the first stage of hematite decomposition for particle size of (63–125) μm.

| Ore | Heating Rate [K/min] | $R^2$ | Model | Rate Constant |
|---|---|---|---|---|
| A | 2 | 0.9910 | F3 | $k = 3.03 \times 10^{21} \frac{1}{min} \times exp(-\frac{670{,}499 \frac{J}{mol}}{RT})$ |
| B | 2 | 0.9899 | F2 | $k = 3.18 \times 10^{18} \frac{1}{min} \times exp(-\frac{587{,}682 \frac{J}{mol}}{RT})$ |
| C | 2 | 0.9773 | F2 | $k = 6.11 \times 10^{23} \frac{1}{min} \times exp(-\frac{592{,}322 \frac{J}{mol}}{RT})$ |

Based on the results, Figure 13 summarises the process of the thermal decomposition of the iron ore used in this study. The illustration shows the decomposition of hematite as only iron-bearing phase (OreA) and the decomposition if both hematite and goethite are present (OreB and OreC) as separated flows. In the decomposition process of OreA, dolomite present in the sample first decomposes, leading to the emission of $CO_2$. For OreB and OreC, goethite decomposition occurs as a two-stage process and is the first identified decomposition process. The first stage of goethite decomposition is assumed to be that of the almost pure goethite particles detected with the Raman spectroscopy. The conversion of goethite to hematite in this stage is found to be controlled by the chemical reaction. In the second stage it is expected that the less accessible goethite, which co-exists with hematite in the same particle, decomposes. The decomposition control shifts from a chemical reaction to diffusion in this stage. The decomposition of the goethite phase is identified by the emission of water vapour. For all ores, the last observed decomposition is the conversion from hematite to magnetite under the release of oxygen at elevated temperatures. Independent of the input ore, the hematite conversion initiates at similar temperatures and is found to be controlled by a chemical reaction for all of the ores.

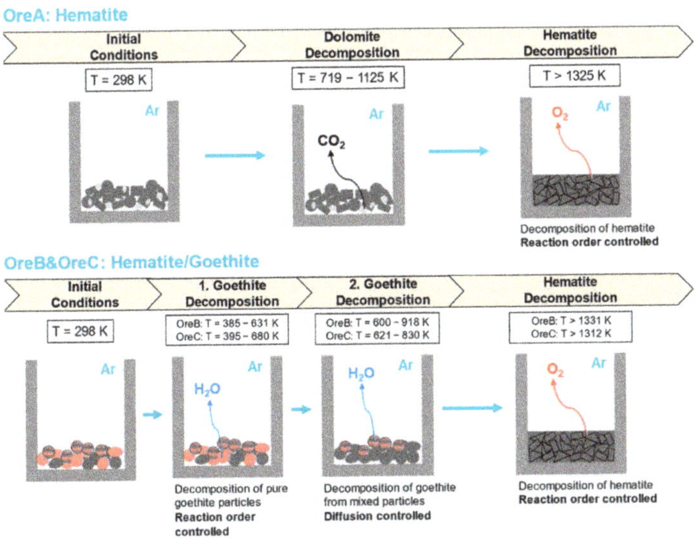


**Figure 13.** Schematic illustration of the thermal decomposition process of varying iron ore fines.

## 5. Conclusions

1. The varying raw iron ores investigated in this study (elaborated in Section 2) differ from one another in morphology, mineralogy and chemical composition. However, after the pre-reduction in an argon atmosphere in the TGA experiments at 1773 K, the morphology of the pre-reduced magnetite phases appeared to be identical.
2. The thermal decomposition of goethite ore occurs as a two-stage process. The mass loss in the individual goethite stages depends on the goethite content in the ore. For the particle size fraction of <63 µm, OreB shows two similar mass loss stages, with an observed mass loss of 0.56–0.67 wt.% in the first and 0.55–0.65 wt.% in the second stage. OreC shows stronger differences with a mass loss of 3.81 wt.%–4.10 wt.% in the first and 0.13–0.20 wt.% in the second stage.
3. The proposed reaction mechanisms for goethite decomposition are a chemical reaction in the first stage and diffusion in the second stage (temperature intervals stated in Table 5). The order of reaction and dimension of diffusion partially varied and is difficult to prove due to the nature of the TGA. Therefore, the investigation of reaction order and diffusion dimension will be within the scope of future research. Due to the significant differences in weight loss in both decomposition stages between OreB and OreC, OreC shows higher activation energies in the first stage (77.86–134.47 kJ/mol) and OreB in the second stage (32.36–44.76 kJ/mol) for a particle size fraction < 63 µm.

4. An increase in particle size has no influence on the reaction mechanisms of goethite thermal decomposition. However, the activation energy remains similar for OreB in the first stage (56.44 → 56.26 kJ/mol) but decreases in the second stage (33.38 → 24.43 kJ/mol). The reason for the change in activation energy is proposed to be the difference in mineralogy between the two particle size fractions. The larger fraction of OreB contains higher amounts of crystal water, possibly leading to an easier release in the second decomposition stage. For OreC, however, the activation energy decreases in both stages (77.86 → 69.90 kJ/mol and 8.46 → 3.32 kJ/mol) with increasing particle size, despite a similar level of crystalline water, arguing for the influence of particle size on the activation energy.
5. Hematite thermal decomposition also occurs as a two-stage process, with the first stage initiating between 1311 and 1419 K and the second stage between 1676 and 1742 K, depending on ore composition and heating rate. A significant amount of decomposition takes place in the first stage. The order of highest average total reduction degree for the size fraction < 63 μm is OreC (14.18–16.49%) > OreB (14.23–14.57%) > OreA (11.82–12.42%). At the same heating rate, an increase in particle size leads to no significant increase in the degree of reduction for all ores.
6. The proposed first stage of hematite decomposition is chemical reaction controlled, while the order of the reaction rate depends on the heating rate and ore type. Again, due the nature of the TGA, the order of the reaction is difficult to prove and will be within the scope of future research. No influence of the particle size on the reaction mechanism was observed. The lowest activation energy for the < 63 μm size fraction was found for OreC (424.31–505.47) kJ/mol. A higher activation energy was determined for OreA (545.47–667.45) kJ/mol and OreB (654.98–831.54) kJ/mol. For OreA and OreC the activation energy increases and for OreB it decreases with an increase in particle size.

**Supplementary Materials:** The following supporting information can be downloaded at: https://www.mdpi.com/article/10.3390/met14111271/s1, Figure S1: XRD pattern of OreA (a), OreB (b) and OreC (c) with the particle size fraction < 63 μm portrait in black and the size fraction (63–125) μm in red; Table S1: Linear regression results of model fitting approach for thermal goethite decomposition (fraction < 63 μm); Table S2: Linear regression results of model fitting approach for hematite thermal decomposition (fraction < 63 μm); Figure S2: XRD patterns of cross-section of OreA, <63 μm, heating rate of 2 K/min, OreB, <63 μm, heating rate of 2 K/min and OreC, <63 μm, heating rate of 5 K/min.

**Author Contributions:** Conceptualization, P.L.; methodology, P.L.; software, P.L.; validation, P.L. and J.C.B.; formal analysis, P.L. and A.M.A.; investigation, P.L., J.C.B. and A.M.A.; resources, Y.Y. and C.Z.; data curation, P.L.; writing—original draft preparation, P.L.; writing—review and editing, P.L., Y.Y., C.Z., S.T.A., N.D., K.M. and J.v.d.S.; visualisation, P.L.; supervision, Y.Y., C.Z., S.T.A., N.D., K.M. and J.v.d.S.; project administration, Y.Y. and C.Z.; funding acquisition, Y.Y. and C.Z. All authors have read and agreed to the published version of the manuscript.

**Funding:** This research was carried out under project number T20009 in the framework of the Research Program of the Materials innovation institute (M2i) (www.m2i.nl) supported by the Dutch government.

**Data Availability Statement:** The original contributions presented in the study are included in the article, further inquiries can be directed to the corresponding author.

**Acknowledgments:** Ruud Hendrikx and Richard Huizenga at the Department of Materials Science and Engineering of the Delft University of Technology are acknowledged for the XRD analysis.

**Conflicts of Interest:** The authors declare no conflicts of interest.

## References

1. Worldsteel Association. *Climate Change and the Production of Iron and Steel*; Worldsteel Association: Brussels, Belgium, 2021.
2. Worldsteel Association. *Sustainable Performance of Steel Industry 2003–2021*; Worldsteel Association: Brussels, Belgium, 2022.
3. Worldsteel Association. *2022 World Steel in Figures*; Worldsteel Association: Brussels, Belgium, 2022.

4. Yang, Y.; Raipala, K.; Holappa, L. Ironmaking. In *Treatise on Process Metallurgy*; Elsevier Ltd.: Amsterdam, The Netherlands, 2014; Volume 3, pp. 2–88. [CrossRef]
5. Biwas, A.K. *Principles of Blast Furnace Ironmaking: Theory and Practice*; Cootha Publishing House: Brisbane, Australia, 1981.
6. Naito, M.; Takeda, K.; Matsui, Y. Ironmaking Technology for the Last 100 Years: Deployment to Advanced Technologies from Introduction of Technological Know-how, and Evolution to Next-generation Process. *ISIJ Int.* **2015**, *55*, 7–35. [CrossRef]
7. Meijer, K.; Guenther, C.; Dry, R.J. *HIsarna Pilot Plant Project*; METEC InSteelCon 2011: Düsseldorf, Germany, 2011.
8. Meijer, K.; Zeilstra, C.; Teerhuis, C.; Ouwehand, M.; van der Stel, J. Developments in Alternative Ironmaking. *Trans. Indian Inst. Met.* **2013**, *66*, 475–481. [CrossRef]
9. Qu, Y.; Yang, Y.; Zou, Z.; Zeilstra, C.; Meijer, K.; Boom, R. Thermal decomposition behaviour of fine iron ore particles. *ISIJ Int.* **2014**, *54*, 2196–2205. [CrossRef]
10. Chen, Z.; Zeilstra, C.; Van Der Stel, J.; Sietsma, J.; Yang, Y. Thermal Decomposition Reaction Kinetics of Hematite Ore. *ISIJ Int.* **2020**, *60*, 65–72. [CrossRef]
11. Hanesch, M. Raman spectroscopy of iron oxides and (oxy)hydroxides at low laser power and possible applications in environmental magnetic studies. *Geophys. J. Int.* **2009**, *177*, 941–948. [CrossRef]
12. Coats, A.W.; Redfern, J.P. Kinetic Parameters from Thermogravimetric Data. *Nature* **1964**, *201*, 68–69. [CrossRef]
13. Pourghahramani, P.; Forssberg, E. Reduction kinetics of mechanically activated hematite concentrate with hydrogen gas using nonisothermal methods. *Thermochim. Acta* **2007**, *454*, 69–77. [CrossRef]
14. Sun, S.S. A Study of Kinetics and Mechanisms of Iron Ore Reduction in Ore/Coal Composites. Ph.D. Thesis, McMaster University, Hamilton, ON, Canada, 1997.
15. Otsuka, R. Recent studies on the decomposition of the dolomite group by thermal analysis. *Thermochim. Acta* **1986**, *100*, 69–80. [CrossRef]
16. Diamandescu, L.; Mihàilà-Tàràbàşanu, D.; Feder, M. On the solid phase transformation goethite → hematite. *Mater. Lett.* **1993**, *17*, 309–311. [CrossRef]
17. Walter, D.; Buxbaum, G.; Laqua, W. The Mechanism of the Thermal Transformation From Goethite to Hematite. *J. Therm. Anal. Calorim.* **2001**, *63*, 733–748. [CrossRef]
18. Xing, L.; Qu, Y.; Wang, C.; Shao, L.; Zou, Z.; Song, W. Kinetic Study on Thermal Decomposition Behavior of Hematite Ore Fines at High Temperature. *Metall. Mater. Trans. B* **2020**, *51*, 395–406. [CrossRef]
19. Sun, Y.; Han, Y.; Wei, X.; Gao, P. Non-isothermal reduction kinetics of oolitic iron ore in ore/coal mixture. *J. Therm. Anal. Calorim.* **2016**, *123*, 703–715. [CrossRef]
20. Pelino, M.; Toro, L.; Petroni, M.; Florindi, A.; Cantalini, C. Study of the kinetics of decomposition of goethitein vacuo and pore structure of product particles. *J. Mater. Sci.* **1989**, *24*, 409–412. [CrossRef]
21. Beuria, P.C.; Biswal, S.K.; Mishra, B.K.; Roy, G.G. Study on kinetics of thermal decomposition of low LOI goethetic hematite iron ore. *Int. J. Min. Sci. Technol.* **2017**, *27*, 1031–1036. [CrossRef]

*metals*

MDPI

*Article*

# Revealing the Surface and In-Depth Operational Performances of Oxygen-Evolving Anode Coatings: A Guideline for the Synthesis of Inert Durable Anodes in Metal Electrowinning from Acid Solutions

**Jovana Bošnjaković [1], Vladimir Panić [1], Maja Stevanović [2], Srecko Stopic [3], Jasmina Stevanović [1], Branimir Grgur [4] and Gavrilo Šekularac [1,*]**

[1] Department of Electrochemistry, Institute of Chemistry, Technology and Metallurgy, National Institute of the Republic of Serbia, University of Belgrade, Njegoševa 12, 11000 Belgrade, Serbia; jovana.bosnjakovic@ihtm.bg.ac.rs (J.B.); panic@ihtm.bg.ac.rs (V.P.); jaca@tmf.bg.ac.rs (J.S.)
[2] Innovation Center of Faculty of Technology and Metallurgy, Karnegijeva 4, 11000 Belgrade, Serbia; majas@tmf.bg.ac.rs
[3] IME Process Metallurgy and Metal Recycling, RWTH Aachen University, Intzestraβe 3, D-52072 Aachen, Germany; sstopic@ime-aachen.de
[4] Faculty of Technology and Metallurgy, University of Belgrade, Karnegijeva 4, 11000 Belgrade, Serbia; bngrgur@tmf.bg.ac.rs
* Correspondence: gavrilo@ihtm.bg.ac.rs

**Abstract:** The electrochemical performances of an oxygen-evolving anode produced by the reactivation of waste Ti substrate by a typical $IrO_2$-$Ta_2O_5$ coating are correlated to the textural (non)uniformities of the coating and its exhaustion state. Coating degradation is considered operational loss of the activity in a metal electrowinning process. It was found that (pseudo)capacitive performances can vary over the coating surface by 20–30% and depend on the type of dynamics of the input perturbation: constant through cyclic voltammetry (CV) or discontinuous time-dependent through electrochemical impedance spectroscopy (EIS). CV-EIS data correlation enabled profiling of the capacitive properties through the depth of a coating and over its surface. The correlation was confirmed by the findings for the analysis of coating activity for an oxygen evolution reaction, finally resulting in the reliable proposition of a mechanism for the operational loss of the anode. It was found that the less compact and thicker coating parts performed better and operated more efficiently, especially at lower operational current densities.

**Keywords:** $IrO_2$-based anodes; operational efficiency in metal electrowinning; electrochemistry of porous electrodes; distributed electrochemical capacitance; anode disfunction

## 1. Introduction

Titanium electrodes with a catalytic coating based on iridium oxide are known to have the best performance for oxygen evolution in industrial electrolysis. These oxygen-evolving anodes (OEAs) are unavoidable in the field of electrogalvanization, water electrolysis, the mining industry, wastewater remediation, impressed current cathodic protection and wastewater electropurification [1–6].

OEAs based on noble metal oxides led to a revolution in electrochemical technologies, owing to their good process stability and excellent catalytic activity [7–9]. However, during the oxygen evolution reaction, simultaneous anodic dissolution of electrochemically active oxide occurs. For example, the much shorter service life of $RuO_2$-based anodes in comparison with $IrO_2$-based anode coatings is due to the much lower anodic stability of $RuO_2$ [10–12]. $RuO_2$ or $IrO_2$ in gas-evolving anodes act as electronic catalysts, while $TiO_2$ and $Ta_2O_5$ preserve the chemical and mechanical stability of the anode oxide coatings [13–15].

Owing to their stabilizing role, valve metal oxides such as $Ta_2O_5$ and $TiO_2$ as well as, for example, $SnO_2$, which can partially operate as a catalyst, reduce the required catalytic powder production costs of the anodes [16–18]. Comninellis and Vercesi [19] showed that a coating with 70 mol% Ir and 30 mol% Ta on Ti has the best activity for oxygen evolution reactions (OERs) and consequently the longest service life in sulfate solutions [20–22].

Coatings for OEAs can be prepared by different methods, such as thermal decomposition of metallic salts, electrodeposition, sol-gel and the Pechini method [23,24]. The anodes used in industries are usually produced with the thermal decomposition method, due to the low cost and simplicity of the process itself. In thermal decomposition, the corresponding precursor salts dissolved in organic or aqueous solutions are applied onto Ti substrate. The coatings obtained under these conditions suffer from inhomogeneous distribution of their properties, with a large number of cracks and the presence of crystalline $IrO_2$ aggregates among the cracks and amorphous $Ta_2O_5$ [25]. The electrocatalytic properties, stability and operational efficiency of the coating are strongly influenced by its surface and microstructural homogeneity degree [26,27]. Consequently, the main challenge is to improve the service life of OEAs in expanding industrial areas, especially at low pH values, moderate electrolyte temperatures and at extremely high current densities [28].

It has been claimed [29] that layered Ir-based coatings of different layer compositions considerably influence the service life of oxygen-evolving electrodes. It follows that the $IrO_2$ in the surface layer (e.g., $IrO_2$-$Ta_2O_5$, 85/15 metal mole ratio) rules the anode service life, which can be considerably improved by 2–2.5 times if coating and support intermediate layers of low $IrO_2$ contents or even without $IrO_2$ are introduced. The compositions of the intermediate layers are based on different transition metal oxides. This stabilizing effect of the Ti support and coating intermediate layers was minimally reported for different testing current densities, and it is by far being considerably more pronounced with respect to other increased service life effects by different coating compositions [30,31].

For this certain benefit of the layered coating, it is required to know how the composition and in-depth distribution of the coating's bulk and surface electrochemical uniformity are correlated for a given application of OEAs. In addition, it is fairly unknown what in-depth changes in activity are taking place in the coating upon anode operation and degradation and how these changes affect the OEA's service life. We present an approach which reveals the changes in activity throughout the $IrO_2$-$Ta_2O_5$ coating upon its destabilization at different current densities of oxygen evolution at low pH levels. The approach is based on analysis of the dynamic electrochemical responses of a coating of an apparent uniform composition to reveal its in-depth operational changes for future modeling of layered structures. Additionally, surface coating uniformity differences are correlated to an in-depth operational coating profile.

## 2. Materials and Methods

### *2.1. Electrode Preparation*

An $IrO_2$-$Ta_2O_5$ porous coating was prepared with the conventional thermal decomposition method [32]. Before applying the coating, the Ti substrate (industrial waste of commercial OEA, 3 cm × 15 cm × 3 mm) was sandblasted, degreased in acetone (for analysis EMSURE® ACS, ISO, Reag. Ph Eur, Supelco, Merck, Kenilworth, NJ, USA) and washed with deionized water (18MΩ, Merck Millipore, Burlington, MA, USA). The subsequent etching was performed in boiling 10% oxalic acid (ACS reagent grade 99%, Sigma Aldrich, St. Louis, MO, USA) for 2 h in order to remove $TiO_2$ and generate a surface with a uniform roughness. The precursors, $H_2IrCl_6xH_2O$ (36.0–44.0% Ir, Sigma Aldrich, USA) and $TaCl_5$ (reagent grade 99%, Sigma Aldrich, USA) at an Ir:Ta mole ratio of 70:30 were dissolved in n-butanol (0.2 M), (ACS reagent 99.4%, Sigma Aldrich, USA) and applied in layers to the Ti substrate with the drop casting method. After application, each layer was dried at a temperature of about 50 °C for 5 min, placed in a drying oven at 100 °C for another 5 min and finally annealed in a high-temperature furnace at 500 °C for 15 min. The procedure was repeated four times, with the final annealing lasting for 2 h. The total amount of oxides applied on the substrate was projected to be 1 mg/cm$^2$. A digital image of the coated electrode is given in Figure 1.

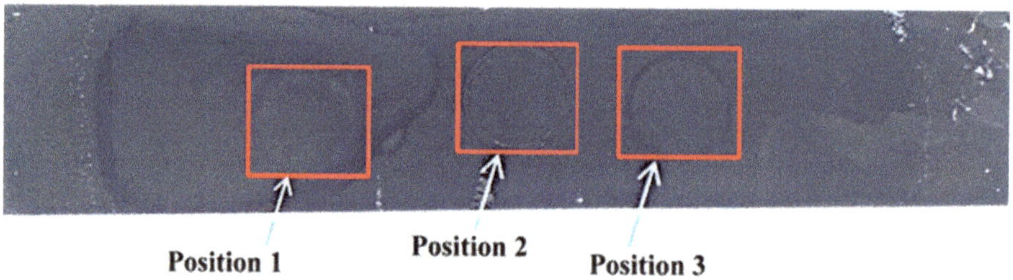


**Figure 1.** Digital image of coated electrode.

*2.2. Electrochemical Measurement*

Electrochemical characterization of the as-prepared and exhausted electrode was performed at three randomly selected positions on the electrode's surface using potentiodynamic polarization, cyclic voltammetry (CV), electrochemical impedance spectroscopy (EIS) and an accelerated stability test (AST). All measurements were performed in a 10% $H_2SO_4$ solution in a high-density polyethylene cell, where a platinum electrode was used as a counter electrode and a saturated calomel electrode (SCE) was the reference. All potentials mentioned are in reference to the SCE scale. The working area of the electrode was 1.33 cm$^2$, whereas the working area of the counter electrode was 20 cm$^2$. All measurements were performed using a Biologic—SP 240 potentiostat (BioLogic, Seyssinet-Pariset, France). For anodic potentiodynamic polarization, the scan rate was 2 mV/s. The cyclic voltammetry curves were recorded within a potential window of water stability at a scan rate of 50 mV/s for the 11th steady state cycle. For the EIS measurements, an AC sine signal amplitude of the root mean square of 10 mV was applied around the open circuit potential in a frequency range of 10 mHz–1 MHz. The impedance spectroscopy data obtained were fitted to an appropriate equivalent electrical circuit using ZView® software (version 3.2b).

The AST was conducted under galvanostatic electrolysis conditions, with a current density of 2.0 or 3.0 A/cm$^2$ at a controlled temperature of 48 °C. The tested electrode was used as the anode, and a platinum wire was used as the cathode. The voltage between the anode and cathode was monitored and recorded. When the voltage reached approximately a 50% increase with respect to the initial voltage of about 6.5 V (set below the potential of the Ti substrate's oxidative breakthrough), the tested electrode was considered to have reached the end of its service life (exhausted electrode). Under the applied conditions, the service life of the prepared electrode was found to be roughly 100 h.

Three randomly selected positions on the electrode's surface were studied to determine the performance uniformity of the coating.

## 3. Results

Figure 2 shows the cyclic voltammograms (CVs) of the prepared $IrO_2$-$Ta_2O_5$ coating at three surface positions on the coating surface before and after the accelerated stability test (AST). The CV responses before and after the AST, as well as all others, were collected from the same positions to check the surface uniformity of the coating's properties. The CVs were of typical shapes and features for noble or transition metal oxide coatings on Ti, with redox peaks in the range from 0.6 to 0.8 V, which is assignable to Ir(III)/Ir(IV) pseudocapacitive redox transitions [32,33]. Figure 2d shows the corresponding voltammetric capacitances, which were assumed to be proportional to the number of active sites in the porous coating accessible via the applied CV dynamic conditions. The pseudocapacitive characteristics were of a moderately homogeneous distribution over the coating's surface; the difference among the three different positions was below 20%. The values at different positions, although randomly chosen, show that positions 1 and 3 reflected the far ends of the characteristics, with position 2 fairly reflecting average characteristics.

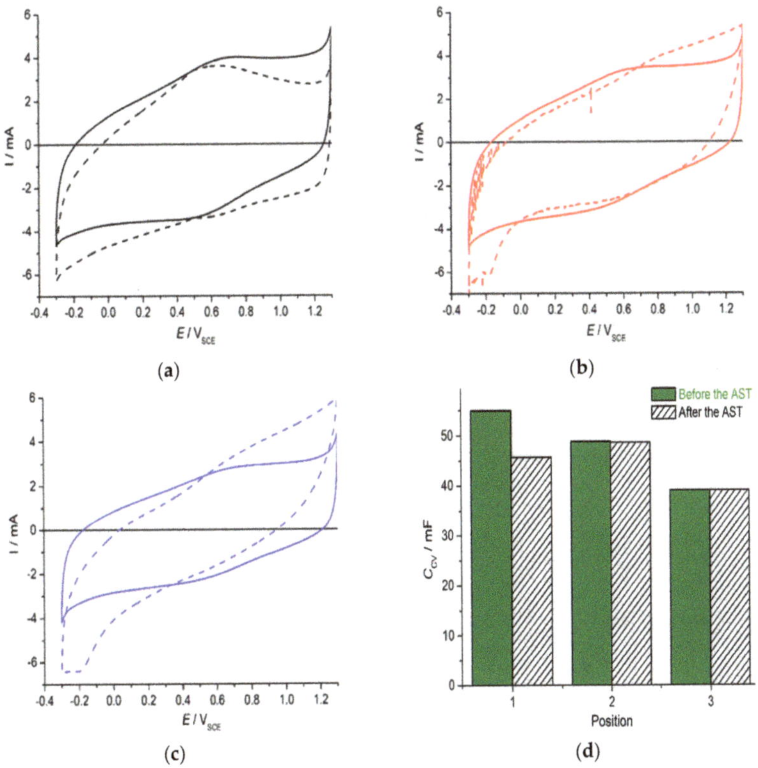


**Figure 2.** Cyclic voltammograms of $IrO_2$-$Ta_2O_5$ coatings on Ti substrate recorded before (full line) and after (dashed line) the accelerated stability test (AST) at positions 1–3 ((**a–c**), respectively). (**d**) Voltammetric capacitances calculated from the curves in (**a–c**).

The number of active sites (coating composition) as well as their distribution and accessibility throughout the porous layer (coating porosity) affect the pseudocapacitive characteristics [34,35]. The former causes the maximal coating performances at the applied composition, while the compact coating surface layer with a small real surface area can be of a lower coating capacitance. Hence, local deviations in the surface coating composition from the nominal composition can produce lower capacitances at the more compact positions (positions 2 and 3), whereas a less compact coating morphology at position 1 can cause a higher capacitance due to the higher number of accessible active sites from the coating's interior.

Another insight into the differences between the surface positions comes from the comparison of the voltammetric capacitances registered after the AST (Figure 2d). The shapes of the CVs with their main features appeared to be unaffected but were clearly tilted with respect to the CVs of the as-prepared coating. This indicates that the coating resistance increased upon application of the AST, being most pronounced for the lowest initial capacitance at position 3. The tilt was not followed by a decrease in voltammetric capacitance, except to some degree for position 1. It can be mentioned that less pronounced surface coating exhaustion during the AST at the initially less compact position 1 did not uncover much of the internal active sites, as was the case for positions 2 and 3. The more compact surface was relatively more exhausted, which allowed the unhindered access to initially loose internal active sites at positions 2 and 3 upon the AST being carried out. The unaffected capacitance values during the AST indicate that active sites, possibly not being distributed along the surface evenly, might have been evenly distributed through the coating's bulk. Consequently, the relative difference between the capacitance values at the three positions after the AST decreased to 12% at most, with respect to 20% for the as-prepared coating state.

To prove and analyze further the proposed operational changes in the coating's structure/performance relationship, the impedance response of the coating was registered to discretize its electrochemical parameters through the coating. Figure 3 shows the Bode plots of the EIS data, with the corresponding impedance and capacitance complex plane plots as insets. EIS data from the high-frequency domain of the as-prepared coating obeyed the capacitive-like response down to several hundreds of mHz in the form of well-resolved capacitive loops, with the sizes following the order indicated by the CV response. At the lowest frequencies, the capacitive response resembled the distributed resistor–capacitor in series, thus indicating the discretized time constants of the layers in the coating interior. This manifested as a tail in the capacitance complex plane plots and a phase angle peak of about 0.1 Hz in the Bode plots, which appeared at somewhat higher frequencies for positions 2 and 3 (noted in the CV analysis as being more compact). This agrees with the CV finding that position 1 had a less compact structure (with wider cracks and pores) with a thin surface layer, whereas the coating interior was of a similar structure to that of positions 2 and 3. This could also explain the registered decrease in CV capacitance upon administering the AST exclusively for position 1. The less compact thin layer was exhausted more and consequently uncovered more of the coating interior for the AST. It follows that, aside from fluctuations in the surface coating compositions among different surface positions, the coating could also be of a distributed thickness. Considering that thermal stress during the coating formation on Ti should increase with the coating thickness, more cracked surface layers could be formed for a thicker coating. The stress should be even more pronounced if the coating is deposited with a layered thermal treatment. The thicker, more cracked coating could produce different performance levels at position 1 through CV and be recognized in the capacitive and distributed elements transition issue for the EIS data.

The coating degradation under the AST almost completely disturbed the clear capacitive response of the as-prepared electrode, with no signs of a high-frequency capacitive loop. The EIS data seemed to obey the distributed elements' behavior in a much wider frequency range, with the appearance of a high-frequency semicircle in the complex plane plot, as is typical for coating failures in the AST [36]. The semicircle features differed for different surface positions, being well resolved at position 1 with a clear phase angle peak at approximately 4 kHz and much like the shoulder extending from position 2 to position 3. The corresponding phase angle feature was moved to lower frequencies. If it is assumed that the semicircle is related to an $IrO_2$-deficient surface coating layer [37], then it can be noted that the layer was best defined and the least compact (appearing at the highest frequency) at what was the least compact surface initially (position 1). The discrete EIS feature caused by coating degradation is seen as resolution of the mid-frequency phase angle region, in addition to the mentioned high- and low-frequency ones. Additional peaks or shoulders, depending on the surface position, appeared. This indicates that the time constants became resolved better upon coating degradation via the AST, since more porous and less compact layers remained.

Bearing in mind the characteristic physicochemical features recognized in the EIS spectra, the equivalent electrical circuit (EEC) which fit the EIS data best was found to be a transmission line of the form $Rs(C_1(R_1(C_2(R_3 \ldots C_{n-1}(R_nC_n)))))$. The resistor $R_s$ bears the parameter of solution resistance, whereas resistor $R_i$ and capacitor $C_i$ represent the coating pore resistance and capacitance of less and less accessible parts of the coating, respectively, from its surface at $i = 1$ toward the least accessible interiors for $i = n$. The increasing n can thus be anticipated to be the coating "penetration depth" of the EIS input signal. EIS data for the EEC are given as lines in Figure 3, for which the relative error of the circuit parameters did not exceed 20%, with chi-squared and sum of squares values below 5% and $10^{-4}$, respectively. For all coating positions before the AST (as-prepared state), the

EEC of n = 5 returned reliable fitting results, while EECs of n = 6 or 7 (only for the EIS data at position 3) were required to quantify the EIS data after the AST, due to the more pronounced distribution of the time constants.

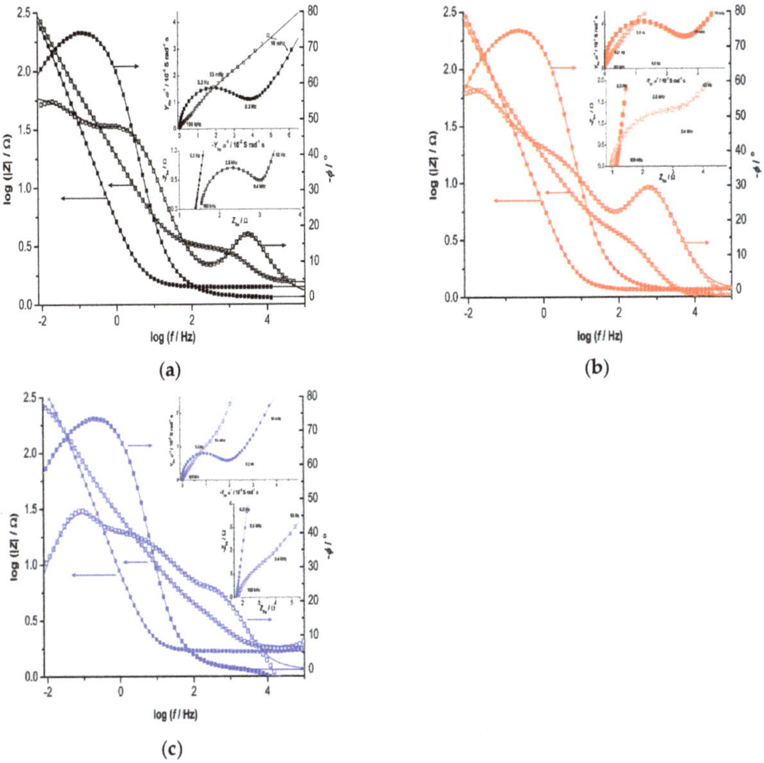


**Figure 3.** Plots of the EIS data (symbols) of the $IrO_2$-$Ta_2O_5$ coating on Ti obtained at the open circuit potential before (full symbols) and after (empty symbols) an accelerated stability test (AST) in 10% $H_2SO_4$ for coating surface positions 1 (**a**), 2 (**b**) and 3 (**c**). The responses of the transmission line's equivalent electrical circuits are shown as lines.

A 3D distribution of the circuit parameter values as a function of the "effective coating depth" (n value) and surface position is presented in Figure 4 for the as-prepared and exhausted coating.

Both the capacitance and pore resistance of the coating changed within two orders of magnitude through its depth, with opposite trends for the as-prepared and exhausted states. The capacitance was the highest and the resistance the lowest for the coating surface layer with respect to the interior layers. Upon degradation under the AST, the capacitance of the surface layer collapsed by three orders of magnitude, while that of the interior layers seemed to be unaffected. Owing to pore widening during the AST, the pore resistance gradually decreased for all layers except on the surface (n = 2), for which the resistance was doubled in comparison with the as-prepared state. Since the coating surface was exhausted the most, access to the active sites beneath was harder to reach for the capacitive response. These findings clearly indicate that the coating's misfunction was caused exclusively by exhaustion of the active sites from the coating surface, while those in the interior stayed intact. This degradation mechanism would cause poor active material efficiency and indicate the requirement for its in-depth distribution to form its surface allocation under adequate coating preparation conditions.

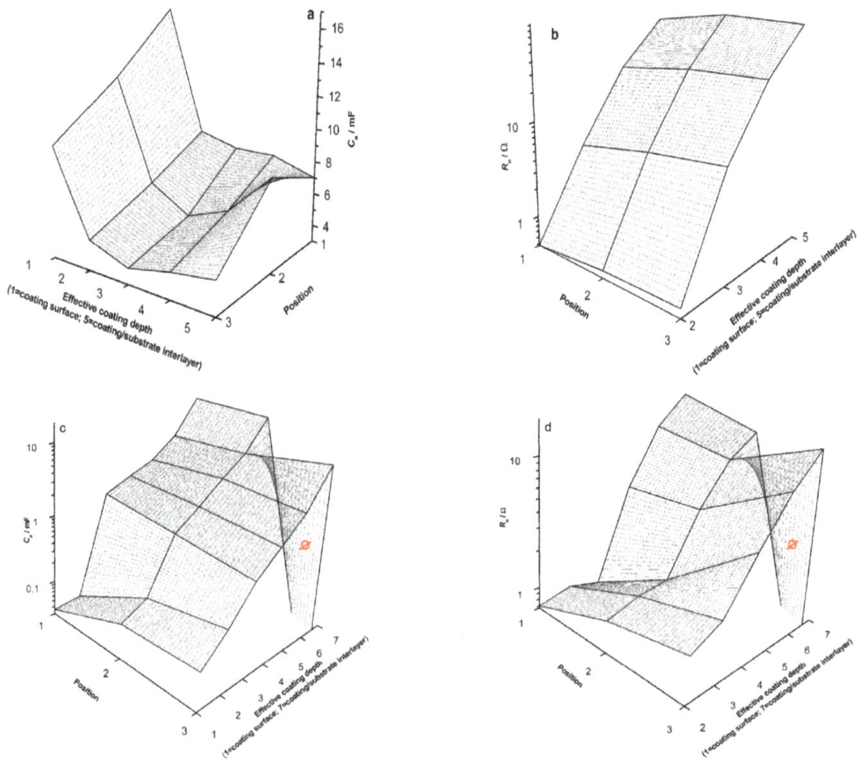


**Figure 4.** Surface (positions 1–3) and in-depth (effective coating depth) distribution of the coating capacitance (**a**,**c**) and pore resistance (**b**,**d**) through the coating layer of the as-prepared (**a**,**b**) and exhausted (**c**,**d**) oxygen-evolving electrode (effective coating depth correlated to the nth order of equivalent transmission line equivalent circuit). Here, 1 = coating surface, n = coating/substrate interface, and ∅ = no data for n = 7 at position 3 after the AST.

Another observation in Figure 4 additionally supports the proposed mechanism and concerns the comparison of surface capacitances (n = 1) among random surface positions. As also found in the CV analysis (Figure 2), the surface capacitance was highest at position 1 for the as-prepared state, which was accessible through the lowest pore resistance (Figure 4a,b). Although this difference in pore resistance with respect to the other two positions became even more pronounced during the AST (Figure 4d, $n \leq 4$), which allowed easier access to the internal capacitance (which was by far the highest for position 1 at n = 4), and the surface capacitance was the lowest during the AST (Figure 4c, $n \leq 2$). This shows that surface exhaustion of the coating's active material was the most pronounced and surface-localized at position 1. However, this behavior during the AST should not be expected, since the AST condition for position 1 was 2 A/cm$^2$, whereas the higher current of 3 A/cm$^2$ was applied to positions 2 and 3. The applied difference was intended to supply data for the influence of the AST-imposed current on the degradation mechanism of provisionally the same coating characteristics. The findings in Figure 4c were the opposite of the expectations, since higher AST currents should produce more vigorous gas evolution and consequently more pronounced inaccessibility to the coating interior (i.e., coating degradation more localized to its surface). According to Figure 4c, a lower AST current apparently affected the degradation at the coating's surface more than hindered accessibility to the coating interior at higher AST currents did. It could be that the lower pore resistance at position 1 allowed much easier evolution of the smaller gas bubbles,

(Their generation rate was below the critical value to allow extending above the critical bubble radius, approaching the pore radius.) which is beneficial for accessibility of the coating interior. These considerations additionally promote the unique requirement of careful planning of the active material's distribution through the coating toward the surface allocation, independent of whether the anode operational conditions would be moderate or vigorous.

Although it followed that an analysis of the CVs (Figure 2) did not supply information about the coating degradation in comparison with the EIS (except for a moderate tilt in the curves), there is other valuable information which can be gained from CV-EIS data correlation. The pseudocapacitive (PC) redox transition processes of noble metal oxides require the insertion (cathodic) or rejection (anodic) of $H^+$ [38], which takes place in a wider potential charging and discharging window applied in CV but not in EIS (rms amplitude of only 10 mV). The voltammetric capacitances from Figure 2d should hence contain a PC contribution of a much higher degree than the overall EIS capacitance through the n branches in Figure 4a,c. If the CV sweep rate as the input charging and discharging perturbation is low enough to penetrate the porous coating, as it would for the lowest input frequencies in EIS, then it can be assumed that a comparison of the CV-EIS capacitances can quantify the PC contribution. Since this contribution is strictly related to the number of active sites, the comparison of the CV-EIS data for the as-received coating at different positions can give insight into the coating composition's uniformity (Table 1).

**Table 1.** Total coating capacitance of as-prepared electrode before accelerated stability test ($\Sigma_{CEIS,ap}$) and after accelerate stability test ($\Sigma C_{EIS,ex}$); the relative contribution of the electrode surface capacitance ($\eta_{Surface}$); and the pseudocapacitive contribution to the total coating capacitance obtained from the difference between the total capacitance calculated from cyclic voltamograms and the total capacitance calculated from EIS ($\delta_{pc} = C_{CV} - \Sigma C_{EIS,ap}$). Calculations for negligible coating pore resistance ($R_n < 1$ and $n \leq 2$) are given in brackets. Data for three different positions are given in the table.

| Parameter | Position 1 | Position 2 | Position 3 |
|---|---|---|---|
| | Before AST | | |
| $\Sigma C_{EIS,ap}$/mF | 48.4 | 38.8 | 27.6 |
| $\Sigma C_{EIS,ap,S}$/mF ($n \leq 3(2)$, $R_n < 10(1)\ \Omega$) | 33.5 (25.5) | 24.4 (19.3) | 18.2 (14.3) |
| | After AST | | |
| $\eta_{Surface}$, % | 69 (53) | 63 (50) | 66 (52) |
| $\delta_{pc}$/% | 14 | 26 | 42 |
| $\Sigma C_{EIS,ex}$/mF | 25.9 | 31.1 | 18.4 |
| $\Sigma\delta_{CEIS,ap\text{-}ex}$/% | 47 | 20 | 33 |
| $\delta_{pc}$/% | 76 | 57 | 112 |

The total coating capacitance through EIS was lower with respect to the voltametric capacitance, but this PC contribution in CV could be considerably different, depending on the coating structure and state. It follows that the PC contribution (δpc) is higher if the total EIS, as well as the voltametric capacitance, are lower (i.e., if the coatings are more compact, thick and of a smaller real surface area). The apparent exception is position 2 after the AST, which indicates that this position was the least compact one upon the AST's application, since position 1, found to be the least compact position in the as-prepared state, was exhausted the most by the AST at a lower current density. (This is also clearly indicated by the highest $\Sigma C_{EIS,ap\text{-}ex}$ value at position 1). Indeed, Figure 4d,c clearly shows that position 2 had the lowest pore resistance and highest capacitance for n = 5 and 6 after the AST.

Different PC contributions caused by the difference in coating structure now appear to be the main cause of the finding that CV was not able to register the difference between the as-prepared and exhausted coating states. The same cause also affected the difference in capacitive performances among different surface positions. In CV, this difference was found to be below 20%, but in the EIS analysis, with a reduced PC contribution, it was found that the difference in total capacitance exceeded 25%, and it was even greater for the surface layers ($\Sigma_{CEIS,ap,S}$). This indicates that the nonuniformity in the coating morphology was more pronounced, especially for the surface texture, than the difference in PC performances affected by the structure of the active sites as well as their number.

The surface layers of the coating were thus found to be the main source of the coating's structural nonuniformity and consequently governed both the basic electrochemical performance and the mechanism of the loss of operational activity. Table 1 presents the values of the surface contribution to the total coating capacitance ($\eta_{Surface}$) down to a certain depth, taking the pore resistance value as the input perturbance limit for penetration. For negligible coating pore resistance ($R_n < 1\ \Omega$ for $n \leq 2$; Figure 3b) approximately 50% of the total available coating capacitance ($\Sigma C_{EIS,ap}$) was easily accessible and created the registered electrochemical performance. This was quite similar at all positions, which indicates that the coating was rather uniform with respect to the in-depth composition. (There was no segregation of the coating components). If the pore resistance increased up to 10 $\Omega$ (which appeared to be its order of magnitude for the exhausted anode), then the relative increase in the surface contribution was only about 15% on average and still similar at all positions. This confirms that the surface led the operation of the anode. Once the surface properties were lost under the AST, there was still roughly 50% of the coating which stayed intact, as was found via CV. However, it was hidden behind a considerably higher resistance, leading to spending higher voltages at the applied AST current densities.

The analyzed coating performances (basic and operational) also affected the coating activity for the oxygen evolution reaction (OER). Figure 5 shows the polarization curves at three different positions on the coating surface for the as-prepared and exhausted states. The data were made to compensate for IR drops to find the true OER activity and Tafel slopes (Figure 5a) and additionally normalized by the total EIS capacitance to reach the true values of the current densities (Figure 5b) (i.e., the rates of OER). The measured values of the Tafel slopes are given in Table 2.

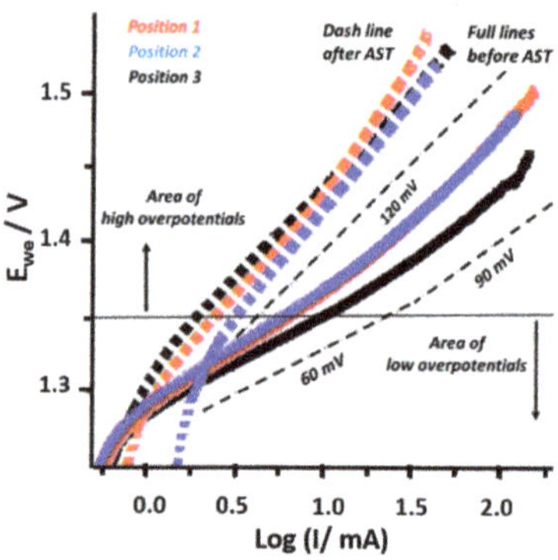


(a)

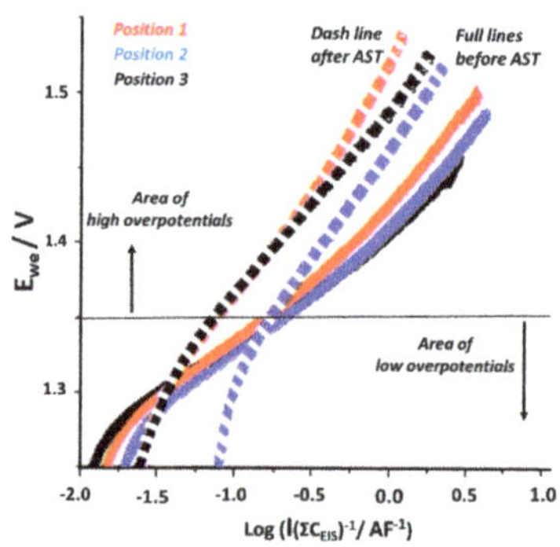


(b)

**Figure 5.** IR drop-corrected (**a**) and normalized (**b**) polarization curves for oxygen evolution reaction on $IrO_2$-$Ta_2O_5$ coating on Ti obtained before (full lines) and after (dash lines) an accelerated stability test (AST) in 10% $H_2SO_4$ at three different coating surface positions.

**Table 2.** Tafel slopes of all three positions before and after accelerated stability test (AST).

| Position | Tafel Slope Before AST (mV dec$^{-1}$) (TS ± SD) | Tafel Slope After AST (mV dec$^{-1}$) (TS ± SD) |
|---|---|---|
| 1 | 64 ± 0.1 Low overpotentials *<br>99 ± 0.5 High overpotentials ** | 123 ± 1 |
| 2 | 78 ± 0.4 Low overpotentials<br>131 ± 0.7 High overpotentials | 143 ± 1 |
| 3 | 73 ± 0.1 Low overpotentials<br>126 ± 0.6 High overpotentials | 143 ± 0.2 |

* Region of the low overpotential was below a potential of 1.35 V vs. SCE. ** Region of the high overpotential was above the potential of 1.35 V vs. SCE.

In accordance with the findings for CV and EIS, the least compact and possibly thickest part of the as-prepared coating (position 1 before the AST) appears to be the most active one (i.e., highest apparent current densities both at low and high OER overpotentials). The more compact positions (positions 2 and 3) showed almost identical activity, despite the differences in capacitive performances. If the total capacitance was taken as a measure of the real surface area to normalize the currents, then all positions were of similar OER activity levels, which indicates that the differences in the apparent currents in Figure 5a were due to differences in the coating texture and not the distributed nonuniform activity. A similar conclusion can also be reached if the Tafel slops are analyzed.

In the as-prepared state, the regions of low and high Tafel slopes were found at the negative (low OER overpotentials) and positive (high OER overpotentials) potentials at approximately 1.35 $V_{SCE}$, respectively. The slopes were close to the theoretical ones (60 and 90 or 120 mV) proposed for the reaction mechanisms of charge transfer or adsorption as rate-determining steps [35]. A unique slope for the exhausted coatings close to or higher than 120 mV was found. This suggests that the most probable slope in the electrocatalytic sense is 120 mV, whereas the halved value of 60 mV at low overpotentials could be due to halving of the activation energy according to the theory of electrocatalysis of porous electrodes [39]. Without delving into the details, it could be mentioned that at low overpotentials, the coating's behavior is porous until saturation of the pores with previously evolved and then dissolved oxygen. Upon saturation and an increase in the amount of evolved oxygen bubbles, which filled the pores at higher overpotentials, the coating's surface participated in the OER, behaving much like a flat electrode. This caused a transition from 60 mV for the porous structure to a "real" slope of 120 mV for a "flat" electrode. Since the exhausted coating had considerably wider pores (considerable decrease in pore resistance; Figure 4b,d), the condition of "electrocatalytic flattening" was not fulfilled, and a unique slope of 120 mV was registered.

These considerations also allow the existence of mixed porous and flat behavior in a high overpotential region if the pores and cracks are wide enough. This behavior can be assigned to the least compact position (position 1) for which the bubble evolution dynamics from the pores allow a great electrocatalytic contribution from the porous structure with a somewhat lower slope of about 90 mV (Figure 5a).

The investigation of the OER kinetics thus supports the findings of the analysis of the coating's electrochemical performances in CV and EIS (i.e., the way the coating texture is reflected by the capacitive and electrocatalytic properties) as well as the coating operation in electrolysis and, consequently, the service life and exhaustion state.

## 4. Conclusions

The standard electrochemical performances of the oxygen-evolving anode $IrO_2$-$Ta_2O_5$/Ti were investigated through capacitive behavior at constant perturbation dynamics through cyclic voltammetry (CV), discontinuous dynamics under electrochemical impedance spectroscopy (EIS) and electrocatalytic activity for oxygen evolution reactions (OERs) via polarization measurements. Performance (non)uniformity was checked over the surface and through the depth of the coating at three different positions on the coating's surface. For in-depth coating performance profiling, the approach of distributed time constants due to time-dependent electrical perturbation was applied. Since the aim was to correlate the texture and uniformity of a coating and its performance to define a pathway for the anode's operational disfunction during the side water electrolysis of the metal electrowinning process, the coating was analyzed in its as-prepared and exhausted states.

More than half of the quantified coating electrochemical activity was found to be related to its surface layers, which also suffered from a considerable loss of activity upon anode disfunction via electrolysis. The performances of the coating interior were affected only negligibly. This finding held for different coating textures, morphologies and electrolysis operational conditions, as it was found to be characteristic of coating nonuniformity due to the thermal procedure of coating formation through drop casting deposition in separate layers. As a guideline for the generation of more efficient coatings, the findings clearly suggest a modified procedure for the distributed composition of deposited layers with top coating layers more rich in the active component $IrO_2$.

Correlation of CV-EIS and OER activity data also indicate that more porous and cracked surface layers, which are more likely to be generated if the coatings are thicker, are more electrochemically efficient, being able to allow the coating to operate more uniformly in electrolysis. In other words, a more porous and thicker coating with a surface enriched by the active component should have a longer service life and better and more uniform material efficiency.

**Author Contributions:** Conceptualization, V.P. and G.Š.; methodology, V.P. and G.Š.; software, J.B.; validation, V.P., G.Š. and B.G.; formal analysis, V.P. and J.B.; investigation, J.B. and G.Š.; resources, J.S. and S.S.; data curation, V.P. and J.B.; writing—original draft preparation, V.P. and J.B.; writing—review and editing, J.B., V.P. and G.Š.; visualization, J.B. and M.S.; supervision, B.G. and G.Š.; project administration, V.P.; funding acquisition, J.S. and S.S. All authors have read and agreed to the published version of the manuscript.

**Funding:** The authors wish to acknowledge the support of the Science Fund of the Republic of Serbia (project number 6666), Renewal of the Waste Oxygen-Evolving anodes from Hydrometallurgy and their improved Activity for Hydrogen Economy, Wastewater and Soil Remediation—OxyRePair.

**Data Availability Statement:** The original contributions presented in the study are included in the article, further inquiries can be directed to the corresponding author.

**Conflicts of Interest:** The authors declare no conflicts of interest.

## References

1. Belal, R.M.; Zayed, M.A.; El-Sherif, R.M.; Abdel Ghany, N.A. Advanced electrochemical degradation of basic yellow 28 textile dye using $IrO_2/Ti$ meshed electrode in different supporting electrolytes. *J. Electrochem. Soc.* **2021**, *882*, 114979. [CrossRef]
2. Ma, R.; Cheng, S.; Zhang, X.; Li, S.; Liu, Z.; Li, X. Oxygen evolution and corrosion behavior of low-$MnO_2$-content Pb-$MnO_2$ composite anodes for metal electrowinning. *Hydrometallurgy* **2016**, *159*, 6–11. [CrossRef]
3. Bebelis, S.; Bouzek, K.; Cornell, A.; Ferreira, M.G.S.; Kelsall, G.H.; Lapicque, F.; de León, C.P.; Rodrigo, M.A.; Walsh, F.C. Highlights during the development of electrochemical engineering. *Chem. Eng. Res. Des.* **2013**, *91*, 1998–2020. [CrossRef]
4. Han, Z.; Zhu, P.; Xu, L.; Kannan, C.S.; Guo, S.; Liu, J.; Koppala, S.; Ju, S. Electrochemical properties of the $IrO_2$-$Ta_2O_5$ coated anodes with Al/Ti and Cu/Ti layered composites substrates. *J. Alloys Compd.* **2018**, *769*, 210–217. [CrossRef]
5. Cao, H.; Zhang, L.; Chen, M. Characteristics of anodic $TiO_2$ nanotube arrays mediated $IrO_2$ Active Anode in the Oxygen Evolution Reaction. *Int. J. Electrochem. Sci.* **2022**, *17*, 220461. [CrossRef]
6. Dondapati, J.S.; Thiruppathi, A.R.; Salverda, A.; Chen, A. Comparison of Pt and $IrO_2$-$Ta_2O_5/Ti$ as a counter electrode in acidic media. *Electrochem. commun.* **2021**, *124*, 106946. [CrossRef]
7. Trasatti, S. *Electrodes of Conductive Metallic Oxides*, 1st ed.; Elsevier Scientific Publishing Company: Amsterdam, The Netherlands, 1980; p. 702.
8. Kawaguchi, K.; Morimitsu, M. Reaction Selectivity of $IrO_2$-Based Nano/Amorphous Hybrid Oxide-Coated Titanium Anodes in Acidic Aqueous Solutions: Oxygen Evolution and Lead Oxide Deposition. *J. Electrochem. Soc.* **2020**, *167*, 133503. [CrossRef]
9. Kasian, O.; Li, T.; Mingers, A.M.; Schweinar, K.; Savan, A.; Ludwig, A.; Mayrhofer, K. Stabilization of an iridium oxygen evolution catalyst by titanium oxides. *J. Phys. Energy* **2021**, *3*, 034006. [CrossRef]
10. Trasatti, S. Electrocatalysis: Understanding the success of DSA®. *Electrochim. Acta* **2000**, *45*, 2377–2385. [CrossRef]

11. Kotz, R.; Stucki, S. Stabilization of $RuO_2$, by $IrO_2$, for anodic oxygen evolution in acid media. *Electrochim. Acta* **1986**, *31*, 1311–1316. [CrossRef]
12. Kamegaya, Y.; Sasaki, K.; Oguri, M.; Asaki, T.; Kobayashi, H.; Mitamura, T. Improved durability of iridium oxide coated titanium anode with interlayers for oxygen evolution at high current densities. *Electrochim. Acta* **1995**, *40*, 889–895. [CrossRef]
13. Tang, D.; Wen, S.; Chen, S. Preparation and characterization for a $RuO_2$ 60%-$TiO_2$ 40% nano-material. *Transc. Met. Heat Treat.* **2000**, *21*, 12–16.
14. Kim, M.; Choi, J.; Lee, W.; Ahn, Y.; Lee, H.; Cho, K.; Lee, J. Performance of Magnéli phase $Ti_4O_7$ and $Ti^{3+}$ self-doped $TiO_2$ as oxygen vacancy-rich titanium oxide anodes: Comparison in terms of treatment efficiency, anodic degradative pathways, and long-term stability. *Appl. Catal. B Environ.* **2023**, *337*, 122993. [CrossRef]
15. Mirseyed, S.F.; Jafarzadeh, K.; Rostamian, A.; Abbasi, H.M.; Ostadhassan, M. A new insight on the mechanisms of corrosion deactivation of a typical Ti/$IrO_2$ + $RuO_2$+$TiO_2$ coating in the presence of $Ta_2O_5$ in chlor-alkali medium. *Corros. Sci.* **2023**, *214*, 111005. [CrossRef]
16. Rosestolato, D.; Neodo, S.; Ferro, S.; Battaglin, G.; Rigato, V.; De Battisti, A. A comparison between structural and electrochemical properties of iridium oxide-based electrocatalysts prepared by sol-gel and reactive sputtering deposition. *J. Electrochem. Soc.* **2014**, *161*, 151–158. [CrossRef]
17. Herrada, R.A.; Acosta-Santoyo, G.; Sepúlveda-Guzmán, S.; Brillas, E.; Sirés, I.; Bustos, E. $IrO_2$-$Ta_2O_5$ | Ti electrodes prepared by electrodeposition from different Ir:Ta ratios for the degradation of polycyclic aromatic hydrocarbons. *Electrochim. Acta* **2018**, *263*, 353–361. [CrossRef]
18. Yang, C.; Shang, S.; Li, X. Oxygen-vacancy-enriched substrate-less SnOx/La-Sb anode for high-performance electrocatalytic oxidation of antibiotics in wastewater. *J. Hazard. Mater.* **2022**, *436*, 129212. [CrossRef]
19. Comninellis, C.; Vercesi, G.P. Characterization of DSA-type oxygen evolving electrodes: Choice of a coating. *J. Appl. Electrochem.* **1991**, *21*, 335–345. [CrossRef]
20. Lee, J.Y.; Kang, D.K.; Lee, K.; Chang, D. An investigation on the electrochemical characteristics of $Ta_2O_5$-$IrO_2$ anodes for the application of electrolysis process. *Mater. Sci. Appl.* **2011**, *2*, 237–243.
21. Xu, W.; Haarberg, G.M.; Seland, F.; Sunde, S.; Ratvik, A.P.; Holmin, S.; Gustavsson, J.; Afvander, Å.; Zimmerman, E.; Åkre, T. The durability of the thermally decomposed $IrO_2$-$Ta_2O_5$ coated titanium anode in a sulfate solution. *Corros. Sci.* **2019**, *150*, 76–90. [CrossRef]
22. Herradaa, R.A.; Rodilb, S.E.; Sepúlveda-Guzmánc, S.; Manríqueza, J.; Exnerd, K.S.; Bustosa, E. Characterization of Ti electrodes electrophoretically coated with $IrO_2$-$Ta_2O_5$ films with different Ir:Ta molar ratios. *J. Alloys Compd.* **2021**, *862*, 158015. [CrossRef]
23. Xu, L.; Xin, Y.; Wang, J. A comparative study on $IrO_2$-$Ta_2O_5$ coated titanium electrodes prepared with different methods. *Electrochim. Acta* **2009**, *54*, 1820–1825. [CrossRef]
24. Herrada, R.A.; Medel, A.; Manríquez, F.; Sirés, I.; Bustos, E. Preparation of $IrO_2$-$Ta_2O_5$ | Ti electrodes by immersion, painting and electrophoretic deposition for the electrochemical removal of hydrocarbons from water. *J. Hazard. Mater.* **2016**, *319*, 102–110. [CrossRef]
25. Kawaguchi, K.; Morimitsu, M. Effects of oxide composition on structure, surface morphology, and oxygen evolution behaviors of $IrO_2$-$Ta_2O_5$/Ti anodes prepared at a high temperature. *Electrochemistry* **2015**, *83*, 256–261. [CrossRef]
26. Xu, W.; Haarberg, G.M.; Sunde, S.; Seland, F.; Ratvik, A.P.; Holmin, S.; Gustavsson, J.; Afvander, Å.; Zimmerman, E.; Åkre, T. Sandblasting effect on performance and durability of Ti based $IrO_2$-$Ta_2O_5$ anode in acidic solutions. *Electrochim. Acta* **2019**, *295*, 204–214. [CrossRef]
27. Guzmána, D.; Dubraya, G.; Aguilarb, C.; Rojasc, P.; Guzmána, A.; Soliz, Á.; Sepúlvedaa, R.; Espinozad, R. Mechanochemical processing of $IrO_2$–$Ta_2O_5$: An alternative route for synthesizing Ir and Ir(Ta)$O_2$ solid solution. *Boletín Soc. Española Cerámica Vidr.* **2021**, *60*, 109–118. [CrossRef]
28. Martelli, G.; Ornelas, R.; Faita, G. Deactivation mechanisms of oxygen evolving anodes at high current densities. *Electrochim. Acta* **1994**, *39*, 1551–1558.
29. Jansen, H.J.; Mackor, A. Anodes with Extended Service Life and Methods for Their Manufacturing. EPO Patent No. 0538955B1, 1998.
30. Yan, Z.; Zhang, H.; Feng, Z.; Tang, M.; Yuan, X.; Tan, Z. Promotion of in situ TiNx interlayer on morphology and electrochemical properties of titanium based $IrO_2$-$Ta_2O_5$ coated anode. *J. Alloys Compd.* **2017**, *708*, 1081–1088. [CrossRef]
31. Liu, B.; Ma, B.; Chen, Y.; Wang, C. Corrosion mechanism of Ti/$IrO_2$-$RuO_2$-$SiO_2$ anode for oxygen evolution in sulfuric acid solution. *Corros. Sci.* **2020**, *170*, 108662. [CrossRef]
32. Liu, Y.; Xu, L.; Xuan, J.; Xin, Y.; Li, Y.; Duan, T.; Liu, F. A comparative study on Ti/$IrO_2$–$Ta_2O_5$ anodes prepared by microwave plasma-assisted sintering and conventional thermal decomposition methods. *J. Mater. Res. Technol.* **2023**, *23*, 1447–1457. [CrossRef]
33. Mehdipour, M.; Tabaian, S.H.; Firoozi, S. Effect of $IrO_2$ crystallinity on electrocatalytic behavior of $IrO_2$–$Ta_2O_5$/MWCNT composite as anodes in chlor-alkali membrane cell. *Ceram. Int.* **2019**, *45*, 19971–19980. [CrossRef]
34. Wang, Y.; Tong, H.; Xu, W. Electrocatalytic activity of Ti/$TiO_2$ electrodes in $H_2SO_4$ solution. *Chin. J. Process. Eng.* **2003**, *3*, 356–360.
35. Ren, Z.; Quan, S.; Gao, J.; Li, W.; Zhu, Y.; Liu, Y.; Chai, B.; Wang, Y. The electrocatalytic activity of $IrO_2$–$Ta_2O_5$ anode materials and electrolyzed oxidizing water preparation and sterilization effect. *RSC Adv.* **2015**, *5*, 8778–8786. [CrossRef]
36. Panić, V.; Dekanski, A.; Mišković-Stanković, V.B.; Milonjić, S.; Nikolić, B. On the deactivation mechanism of $RuO_2$–$TiO_2$/Ti anodes prepared by the sol–gel procedure. *J. Electroanal. Chem.* **2005**, *579*, 67–76. [CrossRef]

37. Conway, B.E. *Electrochemical Supercapacitors: Scientific Fundamentals and Technological Applications*; Springer Science + Business Media, LLC.: New York, NY, USA, 1999; pp. 221–257.
38. Gileadi, E. Simultaneous two-electron transfer in electrode kinetics. *J. Electroanal. Chem.* **2002**, *532*, 181–189. [CrossRef]
39. Chizmadzhevand, Y.A.; Chirkov, Y.G. Porous Electrodes. In *Comprehensive Treatise of Electrchemistry*, 1st ed.; Yeager, E., Bockris, J.O.M., Conway, B.E., Sarangapani, S., Eds.; Springer Science + Business Media: New York, NY, USA, 1983; Volume 6, pp. 317–385.

metals MDPI

Article

# Recovery of Samarium from Waste SmCo Magnets via Selective Precipitation with Ammonium Bicarbonate: Optimization of Process Efficiency

Emircan Uysal [1], Elif Emil-Kaya [2,*], Halide Nur Dursun [1,3], Merve Papakci [1], Sebahattin Gürmen [1] and Bernd Friedrich [4]

1 Department of Metallurgical and Materials Engineering, Istanbul Technical University, 34469 Istanbul, Türkiye; uysale15@itu.edu.tr (E.U.); dursunh17@itu.edu.tr (H.N.D.); papakci19@itu.edu.tr (M.P.); gurmen@itu.edu.tr (S.G.)
2 Department of Materials Science and Engineering, Norwegian University of Science and Technology, 7491 Trondheim, Norway
3 Department of Mineral Processing Engineering, Istanbul Technical University, 34469 Istanbul, Türkiye
4 IME Process Metallurgy and Metal Recycling, RWTH Aachen University, 52062 Aachen, Germany; bfriedrich@ime-aachen.de
* Correspondence: elif.e.kaya@ntnu.no

**Abstract:** SmCo magnets are a common material utilized in advanced technological applications. These magnets contain elevated concentrations of Sm and Co within their structural compositions. Given that both Sm and Co are classified as critical metals, the recycling of these magnets after their operational lifespan is of significant economic and environmental importance. Hydrometallurgical recycling processes represent an effective method for the recycling of these magnets. In this study, a pH-controlled selective precipitation method was developed using two $HNO_3$ solutions with distinct oxidizing properties for the recovery of Sm and Co from end-of-life (EoL) SmCo magnets. In the initial stage of the process, the magnets were leached in a low-oxidizing 2M $HNO_3$ solution, with a 1:30 (*w/v*) solid-to-liquid ratio at 20 °C. This step was undertaken to ensure the dissolution of Fe, thereby creating an environment conducive to its removal from the solution. The leaching experiments resulted in dissolution efficiencies of 95%, 96%, and 96% for Sm, Co, and Fe, respectively. In the second stage, a leaching experiment was performed using 3M $HNO_3$ with a 1:10 solid-to-liquid ratio at 60 °C. Under these conditions, Sm and Co achieved dissolution efficiencies of 99%, while Fe remained undissolved in the solid phase due to hydrolysis at the high temperature, thus increasing the solution purity. In the precipitation process, the pH of both leachates was initially adjusted to 4 to precipitate impurities such as Fe ions. As a result of precipitation at pH 4 in the 2M $HNO_3$ leachate, Fe ions were almost completely removed. This was followed by selective Sm precipitation in the pH range of 5–6.5 using $NH_4HCO_3$. The highest purity of Sm precipitation was achieved when the pH reached 6.5. An increase in Sm precipitation efficiency was observed with increasing pH, with an efficiency of 12.75% at pH 5, which rose to 82.37% at pH 6.5. Furthermore, although the precipitation efficiency of Co increased from 6.25% to 10% within this pH range, no significant difference in the extent of this increase was observed. In the case of the 3M $HNO_3$ leachate, the Sm precipitation efficiency at pH 5 was 44.28%, while at pH 6.5, nearly all of the Sm ions were precipitated. The co-precipitation efficiency at pH 5 was 1.89%, increasing to 36.43% at pH 6.5. This increase in co-precipitation was attributed to the system's Eh value, which reflects the enhanced oxidizing properties of the 3M $HNO_3$ leach solution. The results of the study indicate that as the oxidizing strength of the solution increased, the co-precipitation rate also increased with rising pH.

**Keywords:** SmCo magnets; recycling; selective precipitation; hydrometallurgy

## 1. Introduction

Rare earth elements (REEs) and Co are among the critical raw materials, which are metals with high economic value and supply risk but nevertheless indispensable for use in many advanced technological applications [1]. SmCo magnets, developed by Alden Ray and Karl Strnat in the 1970s, are categorized into two distinct types: $SmCo_5$ and $Sm_2Co_{17}$ [2]. For several decades, SmCo has been recognized as the most thermally durable and energy-dense permanent magnet utilized in various applications. However, during the 1980s, the advent of NdFeB magnets developed by Japanese and American companies, distinguished by their economic efficiency and superior performance, led to a rapid transition from SmCo to NdFeB, particularly because of the challenges associated with cobalt supply at the time. NdFeB magnets have become the alternative in many electronic applications, such as hard disk drives, mobile phones, and electric vehicle motors, due to their lower cost and reduced susceptibility to price fluctuations [3]. Nevertheless, SmCo magnets exhibit distinctive performance characteristics that differentiate them from other types of magnets [4]. Their higher magnetic coercivity and superior temperature resistance make them particularly well-suited for specialized applications, such as vehicle powertrains and aircraft engines [3,4]. Recent research endeavors aim to enhance the performance of SmCo magnets while reducing their cost, with projections indicating an increase in global production in the coming years [5].

The recycling of Sm, Co, and other precious metals contained in end-of-life SmCo (EoL-SmCo) magnets is of significant importance for economic and environmental reasons, particularly in light of the economic importance of these metals and the environmental impact associated with ore production [6]. Hydrometallurgy is a frequently utilized method in the literature for the recycling of SmCo magnets [7–9]. The hydrometallurgical recycling processes of SmCo magnets consist of two steps: (1) leaching and (2) precipitation. Moreover, Sm, Co, and other metals are dissolved; solvent extraction is also used for the separation of metals [10]. Although solvent extraction provides high-efficiency metal separation, it has many disadvantages in terms of the environment and the economy [11]. The selective precipitation method has the potential to effectively separate Sm and Co [12], while simultaneously reducing the economic input of the process and requiring less chemical use. In the study conducted by Wang et al. (2022), waste SmCo magnets were leached with sulfuric acid, and then Fe and Cu ions were removed from the solution with ion resin [12]. Sm and Co remained in the solution; they proposed a process for the separation of Sm and Co from this solution via selective precipitation. In the samarium precipitation process, 98.7% efficiency was obtained by using $Na_2SO_4$ at 80 °C, a 4:1 molar ratio, and 90 min of processing time. The high temperature was determined as an important factor that increased Sm precipitation efficiency. While Sm was selectively precipitated, Co remained in the solution. In another study by Xu et al. (2010), ammonium oxalate was used to precipitate REEs from waste SmCo magnets following sequential leaching with sulfuric acid and sodium persulfate [13]. Initially, sulfuric acid leaching was performed to separate Sm and Co. Subsequent leaching with sodium persulfate was conducted to oxidize Co and Fe, resulting in increased concentrations of Fe and Co in solution and leading to the formation of Co-hydroxide ($Co(OH)_3$) and Fe-hydroxide ($Fe(OH)_3$). After filtration, Co, Fe, and rare earth elements (REEs) were separated. For REE precipitation, ammonium oxalate solution was applied, and REE oxides were obtained through drying and roasting following precipitation. REEs can be precipitated in the form of carbonates at relatively high pH values [14]. Although a variety of carbonate-containing compounds are employed as carbonate precipitating agents, ammonium bicarbonate ($NH_4HCO_3$) has garnered particular attention as an effective precipitating agent, exhibiting both precipitating and pH-increasing effects. In a recent publication, Anawati and Azimi (2022) detailed the stepwise precipitation method and the precipitating effect of ammonium bicarbonate for the separation of REEs from ionic clays originating from South America [15]. In the process, ammonium bicarbonate initially serves as a pH adjuster, selectively removing impurities that precipitate in the form of hydroxide at pH 6. Subsequently, it was reported that ammonium bicarbonate acts as both a carbonate ion source and a pH adjuster in the pH range of 6–7.5, precipitating REEs in the form of carbonate. Ammonium bicarbonate is the preferred precipitating agent in the precipitation stage due to its low cost and environmentally friendly structure. The selective precipitation of Sm with $NH_4HCO_3$ after leaching of SmCo magnets has not, to the best of our knowledge, been attempted previously. $NH_4HCO_3$ has the potential to separate Co and Sm from solution, offering an alternative to ionic liquid or solvent extraction. It has the capacity to reduce process costs and is suitable for large-scale recovery processes. Furthermore, this study demonstrates the impact of solution oxidation degree on precipitation.

This study introduces a pH-controlled selective precipitation method for the recovery of Sm and Co from EoL SmCo magnets using $HNO_3$ solutions with distinct oxidizing properties. By tailoring leaching conditions and employing $NH_4HCO_3$ as both a pH adjuster and carbonate source, the method achieves high Sm precipitation efficiency at pH 6.5 while minimizing Co co-precipitation, offering an innovative, environmentally friendly approach to rare earth element recycling.

## 2. Materials and Methods

### 2.1. Materials

$Sm_2Co_{17}$-type EoL-SmCo magnets were supplied by Vacuumschmelze GmbH & Co. KG (Hanau, Germany). $HNO_3$ (65%, Merck, Darmstadt, Germany) was used for the leaching process of EoL-SmCo magnets. The precipitation processes were conducted using NaOH pellets (Merck, Darmstadt Germany) and $NH_4HCO_3$ (Sigma Aldrich, St. Louis, MO, USA) while Milli-Q water was employed in the leaching, precipitation, and sample preparation procedures. Ethanol and acetone were used at various stages of the process.

### 2.2. Leaching of SmCo Magnets

The leaching process was conducted in a manner consistent with that employed in our previous study [7]. Two distinct leaching procedures were performed. The leaching parameters of both processes are presented in Table 1. The E1 leaching process was conducted with a solution of 2M $HNO_3$, a solid-to-liquid ratio of 1/30 (g/L), a temperature of 20 °C, and a duration of 60 min. The E1 leaching process was carried out using a 2M $HNO_3$ solution with a solid-to-liquid ratio of 1/30 (g/L), at a temperature of 20 °C, and for a duration of 60 min. In contrast, the E2 leaching process was performed with a solution of 3M $HNO_3$, a solid-to-liquid ratio of 1/10 (g/L), a temperature of 60 °C, and 60 min. The impact of the parameters was previously elucidated in our research study. E2 leaching provides a more oxidative solution; therefore, Fe remains in the leach residue. In the precipitation step, it is necessary to remove Fe ions before the precipitation of Sm ions. Otherwise, the presence of Fe ions affects the purity of the obtained precipitate. These two leaching solutions were used to investigate the effect of the removal of Fe ions during the pre-step on the Sm precipitation efficiency and to investigate the selective Sm precipitation behavior in the solution with more oxidizing conditions. Leaching efficiencies were calculated using Equation (1).

$$The\ leaching\ efficiencies\ (\%) = \frac{Metal\ ion\ concentration\ \left(\frac{g}{L}\right) \times Leach\ volume(L)}{Metal\ amount\ of\ specific\ metal\ (\%) \times Initial\ metal\ amount\ (g)} \times 100 \quad (1)$$

**Table 1.** Parameters of the experiments.

| | $HNO_3$ Concentration (M) | Leaching Solid-to-Liquid Ratio (g/L) | Temperature (°C) | Time (min.) |
|---|---|---|---|---|
| E1 | 2 | 1/30 | 20 | 60 |
| E2 | 3 | 1/10 | 60 | 60 |

**Table 1.** *Cont.*

| | Precipitation | |
|---|---|---|
| | Initial pH | Precipitation pH |
| E1P1 | 4 | 5 |
| E1P2 | 4 | 5.5 |
| E1P3 | 4 | 6 |
| E1P4 | 4 | 6.5 |
| E2P1 | 4 | 5 |
| E2P2 | 4 | 5.5 |
| E2P3 | 4 | 6 |
| E2P4 | 4 | 6.5 |

### 2.3. Precipitation

The pH of E1 and E2 leachates was adjusted to 4 using the required 2M NaOH solution. The pH was adjusted to 4 before precipitation, which was employed to remove Fe ions from the E1 solution. This same process was also used to provide a controlled pH increase during the Sm precipitation process with $NH_4HCO_3$ from the E1 and E2 solutions. The precipitation parameters with $NH_4HCO_3$ from solutions adjusted to pH 4 are presented in Table 1. E1P1–E1P4 and E1P1–E2P4 refer to the precipitation tests conducted with E1 and E2 leachates, specifying the pH conditions used in these tests. The initial pH for all tests was 4, while P1, P2, P3, and P4 correspond to pH values of 5, 5.5, 6, and 6.5, respectively. In addition to being a precipitating agent, $NH_4HCO_3$ also increases the pH of the solution. The requisite quantity of 2M $NH_4HCO_3$ was added to the solutions to precipitate the Sm complex within the pH range of 5–6.5, from both solutions. The obtained precipitates were washed 5 times with de-ionized water and dried in a vacuum oven at 50 °C. The experimental scheme of the study is presented in Figure 1. The precipitation efficiencies were calculated using Equation (2).

$$\text{The precipitation efficiencies } (\%) = \frac{\left(\text{Initial Metal ion concentration } \left(\frac{g}{L}\right) - \text{Metal ion concentration after precipitation } \left(\frac{g}{L}\right)\right)}{\text{Initial Metal ion concentration } \left(\frac{g}{L}\right)} \times 100 \quad (2)$$

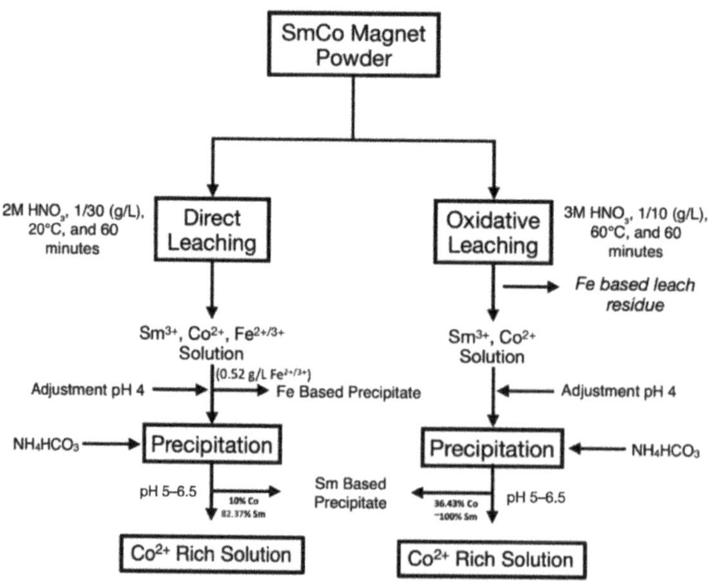


**Figure 1.** Experimental scheme of the study.

### 2.4. Characterization

The precipitation efficiencies of Sm, Co, and Fe were then calculated based on the metal concentrations obtained from the ICP-OES (SPECTRO ARCOS) analysis. SEM/EDS (Scanning Electron Microscope, Thermo Fisher Quattro S, Thermo Fisher Scientific, Brno, Czech Republic) analysis was performed to analyze of precipitates' morphology and elemental distribution.

## 3. Results and Discussion

### 3.1. Characterization of EoL SmCo Magnets

A detailed characterization of waste $Sm_2Co_{17}$ magnets was performed in our previous studies [7,16,17]. The SEM-EDS analysis results indicated that the $Sm_2Co_{17}$ magnet particle distribution was heterogeneous and non-uniform; the particles are primarily composed of Sm (39.68% *w/w*), Co (30.78% *w/w*), and Fe (13.06% *w/w*) metals [16]. Additionally, the EDS results showed the presence of Cu (3.95% *w/w*) and Zr (1.8% *w/w*) [16]. The $Sm_2Co_{17}$ alloy magnets used in high-temperature applications are also $Sm(CoFeCuZr)_z$ (z = 6.0–9.0) quinary alloys [18], which explains the presence of impurities such as Fe, Cu, and Zr alongside Sm and Co in the characterization results. Furthermore, the chemical composition of the magnets was determined using ICP-OES, and the results are presented in Table 2.

**Table 2.** Chemical composition of the magnets.

| Elements | Co | Sm | Fe | Cu | Zr | Nd |
|---|---|---|---|---|---|---|
| Weight, % | 55 | 21 | 15.2 | 5.1 | 2.7 | 0.8 |

The XRD analysis showed that $Sm_2Co_{17}$ was the dominant phase in the structure of the magnets and the FeCo phase was also present, while dynamic particle analysis revealed that 90.3% of the particles were smaller than 96.25 μm [17].

### 3.2. Leaching Reactions

In a previous study involving oxidative leaching of SmCo magnets to remove Fe, leaching reactions were carried out, and the results were examined in detail [7]. As a result of testing numerous parameters, the optimal oxidative leaching conditions were determined through Taguchi plots and ANOVA analysis to be a 3M $HNO_3$, a solid-to-liquid ratio of 1/10, and a process temperature of 60 °C. Leaching efficiencies obtained with E1 and E2 reactions are presented in Figure 2a; also, the metal ion concentrations of leachate solutions are given in Figure 2b. The leaching reaction was performed in 2 M $HNO_3$ at a 1/30 (*w/v*) solid–liquid ratio and 20 °C (E1). The final solution pH was 0.2. In the leaching reaction with 3 M $HNO_3$, a 1/10 solid–liquid ratio, and 60 °C (E2), Sm and Co were efficiently leached (99%), while Fe remained in the residue due to hydrolysis at higher temperatures [7]. The final solution pH was ~2.67, influenced by the high solid–liquid ratio. It is thought that the increase in pH may have favored the transfer of $Fe^{3+}$ ions into the precipitate phase as goethite. The Pourbaix diagram of Sm, Co, and Fe is shown in Figure 3a. In addition, the heat map plotted according to the stability of the $FeHO_2$ structure is shown in Figure 3b. It was observed that after pH around 2, Fe transforms from ionic form to solid form, while $Sm^{2+}$ and $Co^{2+}$ remain in the solution. In addition, the heat map showed that the $\Delta G_{pbx}$ (Gibbs free energy) value (eV/atom) for $FeHO_2$ was 0.13 eV/atom at around pH 2 under oxidizing conditions. This shows that the structure maintains stability under these conditions and Fe remains in the residue.

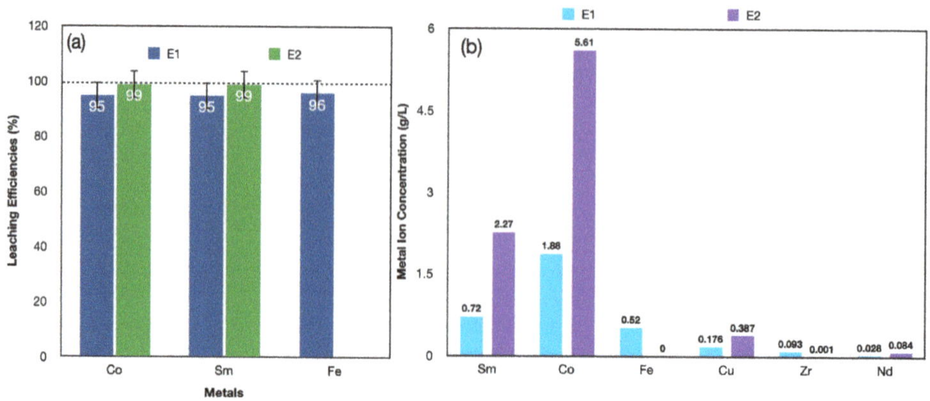


**Figure 2.** Leaching efficiencies (**a**) and metal ion concentration of leachates (**b**).

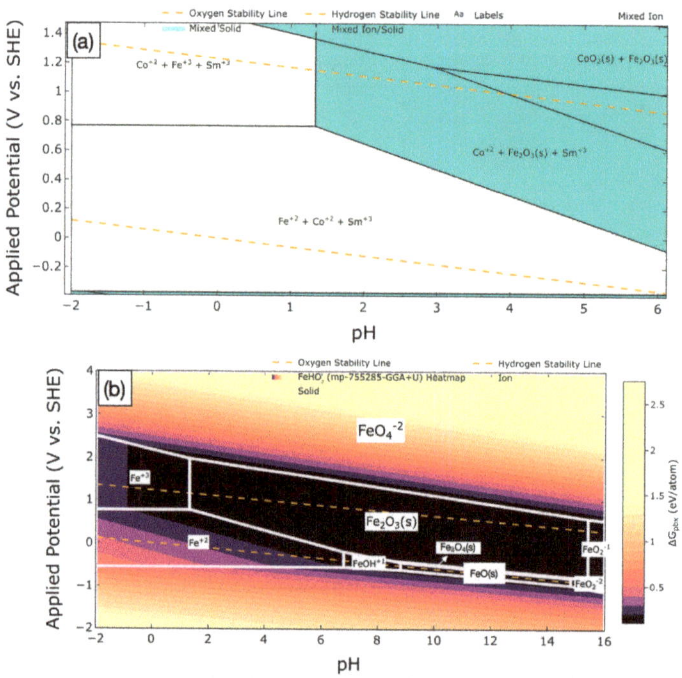


**Figure 3.** The Pourbaix diagram of Sm, Co, and Fe (**a**) and the heat map of $FeHO_2$ GGA + U) (**b**) (drawn with The Materials Project) (Adapt from [19–21]).

### *3.3. Precipitation*

The presence of Fe ions in the solution has an impact on the purity of the resulting precipitate. The precipitation of impurity ions, including $Fe^{3+}$, $Al^{3+}$, and $Ca^{2+}$, at pH values greater than 3–4 facilitates the purification of the leach solution from impurities [22]. In this study, to remove impurity ions in the solution, the pH value of the leaching solutions was increased to 4 before selective REE precipitation with $NH_4HCO_3$, and in particular, the high concentration of Fe ions, obtained after the E1 leaching reaction, was removed from the solution via precipitation. In Figure 4, the precipitation efficiencies of Sm, Co, and Fe ions at pH 4 from the E1 leaching solution are presented. At a pH of 4, the Fe ions were almost completely removed from the solution, while the loss of Sm and Co was calculated to be 0.44% and 3.94%, respectively. While Fe ions and other impurity ions precipitate, even if the pH range of the precipitation does not correspond to that of the REEs ions, these ions precipitate together with the impurities or can adsorb onto the surface of precipitated impurity complexes [23].

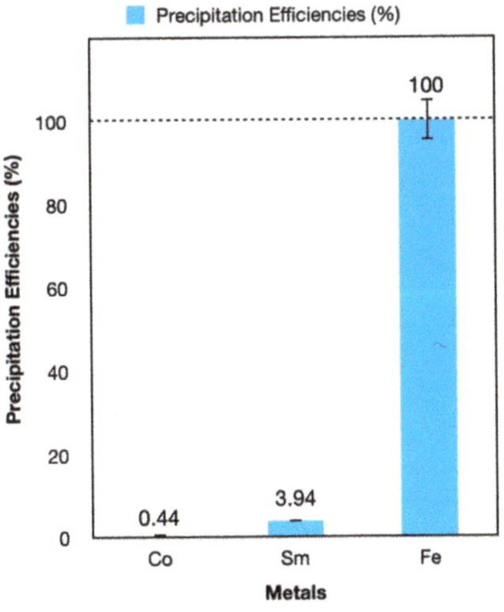


**Figure 4.** Precipitation efficiencies from E1 leachate at pH 4.

REEs can be easily precipitated by carbonate at high pH values, and carbonates are ideal precipitants due to their easy availability, low cost, and easy dissolution in mild acids for subsequent purification; however, since precipitation is difficult in acidic environments, a neutralization process is applied to the leach solution before REE precipitation, and during this process, impurities such as $Fe^{3+}$ and $Ca^{2+}$ are primarily precipitated, which may lead to the loss of REEs through adsorption and co-precipitation with oxides or complexes [14]. To precipitate REE carbonate complexes under mild acid conditions and remove impurities, the pH value in both solutions was adjusted to 4 and then precipitated with $NH_4HCO_3$. It is appropriate to utilize $NH_4HCO_3$ as a precipitating agent for REEs. In addition to providing a carbonate source for complexation with REEs, it serves as a pH adjuster compared to other agents that offer carbonate sources [15,24]. REEs precipitate as REE carbonates ($RE_2(CO_3)_3$) with Equation (3) in the pH range of 7–8 [15]. However, the cobalt carbonate precipitate formation reaction occurs in the pH range of 6.5–8 [24]. To achieve selective precipitation of Co and Sm, precipitation was conducted within a pH range of 5–6.5. Sm and Co precipitation efficiencies from solution E1 in the pH range of 5–6.5 are given in Figure 5a. An increase in pH value from 5 to 6.5 resulted in a notable enhancement in the precipitation efficiency of Sm, with a significant rise from 12.75% to 82.37%. Nevertheless, during the selective precipitation process, some Co was also precipitated via co-precipitation or the adsorption of the formed Sm complex precipitates. Increasing the pH from 5 to 6.5 in solution E1 increased the Co-precipitation rate from 6.25% to 10%. Although this increase was attributable to the approach of the precipitation pH range of Co, no notable distinction was discerned between pH 5 and 6.5. Sm and Co precipitation efficiencies with $NH_4HCO_3$ from E2 solution (in the pH 5–6.5 range) are given in Figure 5b. The precipitation efficiency of the Sm carbonate was found to be 44.28% at pH 5, with the precipitation of all Sm ions in the solution occurring at pH 6.5. The Co-precipitation rate was observed to be 1.89% at pH 5 and 36.43% at pH 6.5. This considerable enhancement in the rate of Co-precipitation is contingent upon the Eh value of the system. At elevated temperatures, the solution obtained through leaching with $HNO_3$ (E2) exhibits enhanced oxidizing properties [7]. It was observed that increasing the oxidizing properties of the solution increased the Co-precipitation rate with increasing pH.

$$2REE^{3+}{}_{(aq)} + 6NH_4HCO_{3(s)} \rightarrow REE_2(CO_3)_{3\,(s)} \downarrow + 6NH_4^+{}_{(aq)} + 3H_2O_{(l)} + 3CO_{2(g)} \quad (3)$$

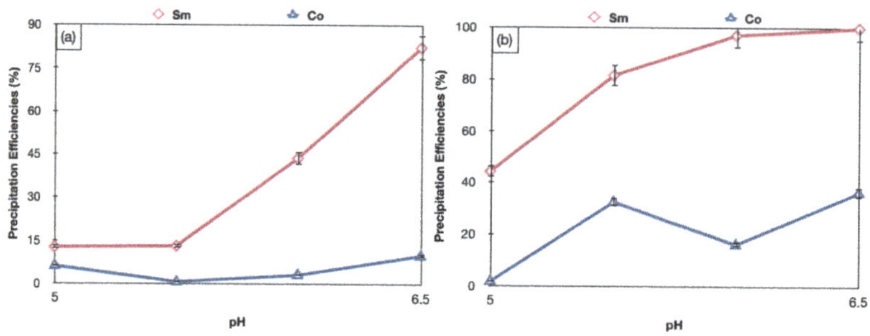


**Figure 5.** Precipitation efficiencies from E1 (**a**) and E2 (**b**) leachate solutions with $NH_4HCO_3$ in the 5–6.5 pH range.

Figure 6 illustrates the SEM images of the precipitates that were formed from the E1 and E2 solutions at pH values of 5 and 6.5, respectively. It was observed that the precipitates exhibited an irregular shape and agglomeration, similar to that observed in the particles obtained as a result of REE precipitation with $NH_4HCO_3$ in the literature. It was observed that the precipitates obtained from the solution with more oxidative conditions exhibited a particle size distribution that was smaller.

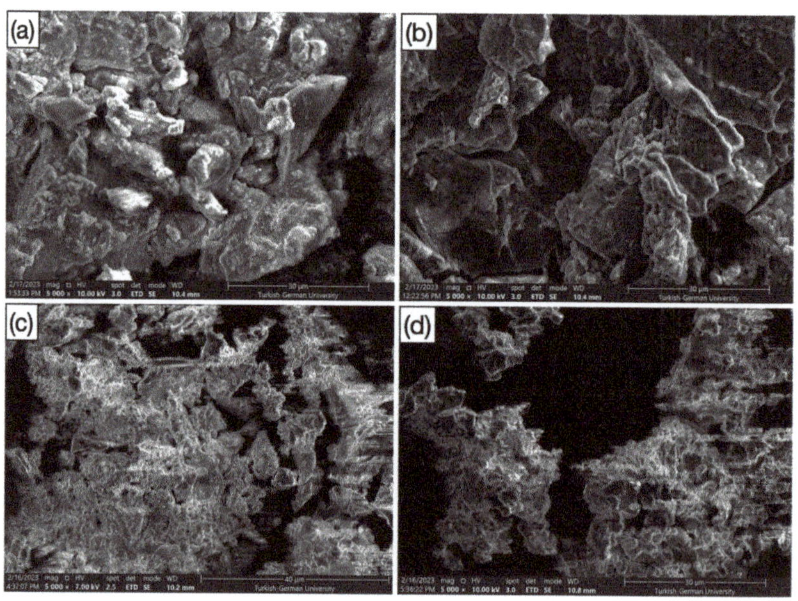


**Figure 6.** SEM images of the precipitate: from E1 solution at pH 5 (**a**) and pH 6.5 (**b**) and E2 solution at pH 5 (**c**) and pH 6.5 (**d**).

EDS maps of the precipitate obtained as a result of precipitation from solution E1 at pH 6.5 are given in Figure 7. It was observed that the distribution of Sm, Co, C, and O in the precipitate was homogeneous, occurring in the same regions. This indicates that Co may either co-precipitate during the precipitation of Sm with carbonate or precipitate by adsorption on the surface of the formed Sm carbonate complexes.

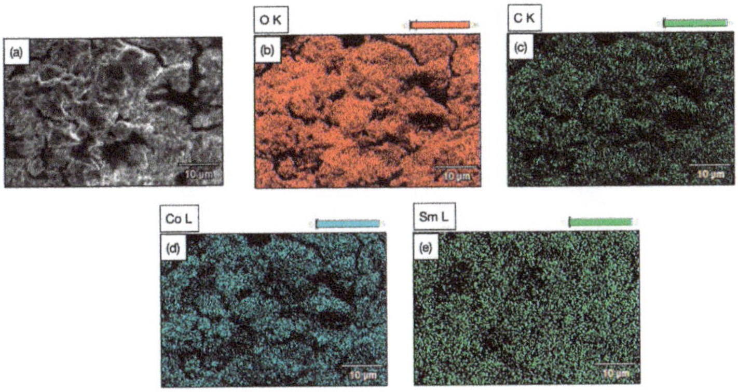


**Figure 7.** SEM images of the precipitate obtained at pH 6.5 from E1 (**a**), and EDS maps of O (**b**), C (**c**), Sm (**d**), and Co (**e**).

EDS results of 3 different points of the precipitate obtained from the E1 solution at pH 6.5 are given in Figure 8. It was observed that three distinct points exhibited a Sm concentration exceeding 60% by weight, with the predominant element in the precipitate being Sm. Furthermore, it was established that the concentration of Co remained below 3% by weight at three distinct points. Additionally, it was determined that Co co-precipitated or adsorbed to Sm complexes during precipitation. It is postulated that the observed Cu concentration (average 3.7% by weight) is derived from the SEM-EDS substrate. Moreover, no Fe was detected in the precipitate, and it was observed that the Fe was effectively removed before carbonate precipitation.

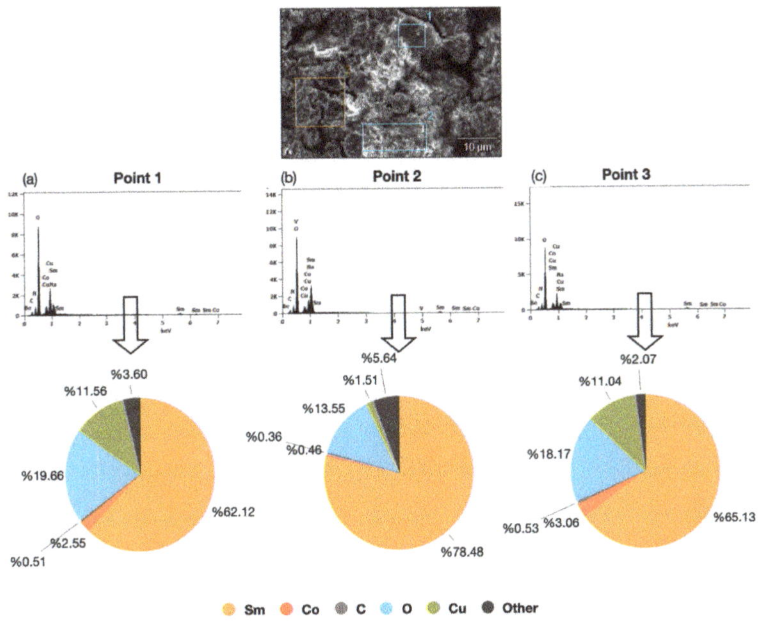


**Figure 8.** EDS results of 3 different points: point 1 (**a**), 2 (**b**), and 3 (**c**).

### 3.4. Comparison of Different Routes

This study was compared with our previous research in terms of process efficiency, the number of steps, environmental impact, advantages, and challenges. In the first route, acid baking with $HNO_3$ was carried out, followed by the separation of REEs through water leaching [17]. In the second route, acid baking with $H_2SO_4$ was performed, followed by REE separation via water leaching [16]. Table 3 presents a comparison of three different routes.

**Table 3.** Comparison of the three different routes.

| Aspect | Acid Baking with $HNO_3$ and Water Leaching | Acid Baking with $H_2SO_4$ and Water Leaching | Oxidative Leaching and Selective Precipitation |
|---|---|---|---|
| Process Step | 1. Mixture with $HNO_3$,<br>2. Selective oxidation at 250 °C<br>3. Water leaching | 1. Mixture with $H_2SO_4$,<br>2. Selective oxidation at 800 °C<br>3. Water leaching | 1. Oxidative leaching<br>2. Iron precipitation<br>3. Selective REE precipitation |
| Efficiency | >95% Sm extraction, <1% Co/Fe dissolution, and 99.4% purity of REEs in leach liquor | 65% Sm extraction; Co/Fe oxides separated; minimal solubility of Fe, Co, and Cu (~2%). | 99% Sm/Co dissolution at 60 °C; selective Sm precipitation (82.37–100%) at pH 6.5 with minimal Co co-precipitation (10–36%). |
| Environmental Impact | NOx gas emission during acid baking; proposed reactor to address emissions. Lower selective oxidation temperature | SOx gas emissions recycled for $H_2SO_4$ production. Higher selective oxidation temperature | Lower-temperature leaching reduces energy demands; careful pH control minimizes reagent use and impurity loss |
| Advantages | High Sm purity and low impurity dissolution; scalable with improved reactor design | Cost-effective, industrially scalable process | High dissolution and purification efficiencies; selective Sm recovery with minimal Co loss; scalable and versatile |
| Challenges | NOx management and reactor scalability | Sm-oxysulfate formation limits extraction efficiency but offers reuse potential | Co co-precipitation increases at higher pH; small amounts of Sm loss in iron precipitation |

## 4. Conclusions

This study demonstrates the effectiveness of an optimized hydrometallurgical process for selectively recovering Sm and Co from EoL SmCo magnets with high dissolution and purification efficiencies. Using 2M $HNO_3$ at 20 °C with a 1:30 ($w/v$) solid–liquid ratio, dissolution efficiencies of 95% for Sm, 96% for Co, and 96% for Fe were achieved. In contrast, leaching with 3M $HNO_3$ at 60 °C and a 1:10 ratio resulted in 99% dissolution efficiency for Sm and Co, while Fe remained undissolved due to hydrolysis, enabling effective impurity removal during purification. In the E1 solution, at pH 4, nearly all Fe was removed, minimizing Sm (0.44%) and Co (3.94%) losses. The use of $NH_4HCO_3$ in the pH 5–6.5 range enabled selective precipitation from the purified leach solution, achieving 82.37% Sm precipitation at pH 6.5 while limiting Co precipitation to 10%. In the E2 solution, the selective precipitation efficiency of Sm increased from 44.28% at pH 5 to 100% at pH 6.5, while Co co-precipitation increased from 1.89% to 36.43%. These results indicate that despite the increased oxidative conditions in E2, high-purity Sm recovery is achievable with controlled pH adjustment. This scalable method offers an efficient and selective approach to rare earth element recovery from waste materials. Further optimization of selective precipitation parameters may extend its applicability to other REEs, advancing sustainable recycling solutions.

**Author Contributions:** E.U.: validation, investigation, formal analysis, writing—original draft, writing—review and editing, and visualization. E.E.-K.: methodology, validation, investigation, formal analysis, writing—original draft, writing—review and editing, and conceptualization. H.N.D.: validation, formal analysis, writing—original draft, and writing—review and editing. M.P.: validation, methodology, and investigation. S.G.: validation, resources, and supervision. B.F.: validation, resources, supervision, and project administration. All authors have read and agreed to the published version of the manuscript.

**Funding:** The project (Waste2Space, Developing High-Value Aerospace Alloys through Waste Valorization) has received funding from the European Union's MSCA-IF program under Grant Agreement N° 101154933.

**Data Availability Statement:** The raw data supporting the conclusions of this article will be made available by the authors on request.

**Acknowledgments:** The project (Waste2Space, Developing High-Value Aerospace Alloys through Waste Valorization) has received funding from the European Union's MSCA-IF program under Grant Agreement N° 101154933.

**Conflicts of Interest:** The authors declare that they have no known competing financial interest or personal relationships that could have appeared to have influenced the work reported in this paper.

## References

1. Koese, M.; Parzer, M.; Sprecher, B.; Kleijn, R. Self-Sufficiency of the European Union in Critical Raw Materials for E-Mobility. *Resour. Conserv. Recycl.* **2025**, *212*, 108009. [CrossRef]
2. Orefice, M.; Audoor, H.; Li, Z.; Binnemans, K. Solvometallurgical Route for the Recovery of Sm, Co, Cu and Fe from SmCo Permanent Magnets. *Sep. Purif. Technol.* **2019**, *219*, 281–289. [CrossRef]
3. Bailey, G.; Orefice, M.; Sprecher, B.; Önal, M.A.R.; Herraiz, E.; Dewulf, W.; Van Acker, K. Life Cycle Inventory of Samarium-Cobalt Permanent Magnets, Compared to Neodymium-Iron-Boron as Used in Electric Vehicles. *J. Clean. Prod.* **2021**, *286*, 125294. [CrossRef]
4. Su, X.; Wang, Y.; Guo, X.; Dong, Y.; Gao, Y.; Sun, X. Recovery of Sm(III), Co(II) and Cu(II) from Waste SmCo Magnet by Ionic Liquid-Based Selective Precipitation Process. *Waste Manag.* **2018**, *78*, 992–1000. [CrossRef]
5. Sinha, M.K.; Pramanik, S.; Kumari, A.; Sahu, S.K.; Prasad, L.B.; Jha, M.K.; Yoo, K.; Pandey, B.D. Recovery of Value Added Products of Sm and Co from Waste SmCo Magnet by Hydrometallurgical Route. *Sep. Purif. Technol.* **2017**, *179*, 1–12. [CrossRef]
6. Li, X.; Li, Z.; Orefice, M.; Binnemans, K. Metal Recovery from Spent Samarium–Cobalt Magnets Using a Trichloride Ionic Liquid. *ACS Sustain. Chem. Eng.* **2019**, *7*, 2578–2584. [CrossRef]
7. Emil-Kaya, E.; Papakci, M.; Friedrich, B. Recycling of SmCo Magnets by Removal of Iron via Oxidative Leaching. *Min. Metall. Explor.* **2024**, *41*, 479–486. [CrossRef]
8. Jha, M.K.; Kumari, A.; Panda, R.; Rajesh Kumar, J.; Yoo, K.; Lee, J.Y. Review on Hydrometallurgical Recovery of Rare Earth Metals. *Hydrometallurgy* **2016**, *165*, 2–26. [CrossRef]
9. Chen, Q.; Ni, S.; Ai, G.; Zhang, T.; Sun, X. A Recovery Strategy of Sm, Co for Waste SmCo Magnets by Fatty Acid Based Ionic Liquids. *Miner. Eng.* **2020**, *158*, 106581. [CrossRef]
10. Zhang, S.; Ni, S.; Zeng, Z.; Mo, D.; Huang, B.; Sun, X. A Sustainable Separation Strategy for Recovering Sm/Co from SmCo Magnets with Fatty Acid and Primary Amine. *J. Mol. Liq.* **2023**, *392*, 123490. [CrossRef]
11. Vera, M.L.; Torres, W.R.; Galli, C.I.; Chagnes, A.; Flexer, V. Environmental Impact of Direct Lithium Extraction from Brines. *Nat. Rev. Earth Environ.* **2023**, *4*, 149–165. [CrossRef]
12. Wang, J.-Z.; Hsieh, Y.-H.; Tang, Y.-C.; Shen, Y.-H. Separation of Cobalt, Samarium, Iron, and Copper in the Leaching Solution of Scrap Magnets. *Metals* **2022**, *13*, 90. [CrossRef]
13. Xu, T.; Zhang, X.; Lin, Z.; Lü, B.; Ma, C.; Gao, X. Recovery of Rare Earth and Cobalt from Co-Based Magnetic Scraps. *J. Rare Earths* **2010**, *28*, 485–488. [CrossRef]
14. Han, K.N. Characteristics of Precipitation of Rare Earth Elements with Various Precipitants. *Minerals* **2020**, *10*, 178. [CrossRef]
15. Anawati, J.; Azimi, G. Separation of Rare Earth Elements from a South American Ionic Clay Lixivium by Sequential Precipitation. *Hydrometallurgy* **2022**, *213*, 105946. [CrossRef]
16. Papakci, M.; Emil-Kaya, E.; Stopic, S.; Gurmen, S.; Friedrich, B. Recovery of Valuable Metals from SmCo Magnets through Sulfation, Selective Oxidation, and Water Leaching. *Sep. Sci. Technol.* **2024**, *59*, 1241–1254. [CrossRef]
17. Emil-Kaya, E.; Lu, X.; Friedrich, B. Recovery of Samarium and Cobalt/Iron Oxide from SmCo Magnets through Acid Baking and Water Leaching. *J. Mater. Cycles Waste Manag.* **2024**, *26*, 3905–3916. [CrossRef]
18. Wang, C.; Shen, P.; Fang, Y.; Wang, S.; Li, Q.; Wang, L.; Li, W.; Zhu, M. Cellular Microstructure Modification and High Temperature Performance Enhancement for Sm2Co17-Based Magnets with Different Zr Contents. *J. Mater. Sci. Technol.* **2022**, *120*, 8–14. [CrossRef]
19. Singh, A.K.; Zhou, L.; Shinde, A.; Suram, S.K.; Montoya, J.H.; Winston, D.; Gregoire, J.M.; Persson, K.A. Electrochemical Stability of Metastable Materials. *Chem. Mater.* **2017**, *29*, 10159–10167. [CrossRef]
20. Patel, A.M.; Nørskov, J.K.; Persson, K.A.; Montoya, J.H. Efficient Pourbaix Diagrams of Many-Element Compounds. *Phys. Chem. Chem. Phys.* **2019**, *21*, 25323–25327. [CrossRef]

21. Persson, K.A.; Waldwick, B.; Lazic, P.; Ceder, G. Prediction of Solid-Aqueous Equilibria: Scheme to Combine First-Principles Calculations of Solids with Experimental Aqueous States. *Phys. Rev. B* **2012**, *85*, 235438. [CrossRef]
22. Han, K.N.; Kim, R. Thermodynamic Analysis of Precipitation Characteristics of Rare Earth Elements with Sulfate in Comparison with Other Common Precipitants. *Minerals* **2021**, *11*, 670. [CrossRef]
23. Zhang, J.; Mani, R.; Louhi-Kultanen, M. Process Monitoring of Cobalt Carbonate Precipitation by Reactions between Cobalt Sulfate and Sodium Carbonate Solutions to Control Product Morphology and Purity. *Hydrometallurgy* **2024**, *224*, 106232. [CrossRef]
24. Kim, P.; Anderko, A.; Navrotsky, A.; Riman, R. Trends in Structure and Thermodynamic Properties of Normal Rare Earth Carbonates and Rare Earth Hydroxycarbonates. *Minerals* **2018**, *8*, 106. [CrossRef]

metals

MDPI

Article

# Kinetics of Sphalerite Leaching by $MnO_2$-KI Oxidation System in Sulfuric Acid

**Mladen Bugarčić, Aleksandar Jovanović *, Dimitrije Anđić, Ivana Jelić, Milica Miletić, Branislav Marković and Miroslav Sokić**

Institute for Technology of Nuclear and Other Mineral Raw Materials, Boulevard Franše d'Eperea 86, 11000 Belgrade, Serbia; m.bugarcic@itnms.ac.rs (M.B.); d.andjic@itnms.ac.rs (D.A.); i.jelic@itnms.ac.rs (I.J.); m.miletic@itnms.ac.rs (M.M.); b.markovic@itnms.ac.rs (B.M.); m.sokic@itnms.ac.rs (M.S.)

* Correspondence: a.jovanovic@itnms.ac.rs; Tel.: +381-11-3691-722

**Abstract:** Efficient extraction of zinc from polymetallic concentrates is crucial for the metallurgical industry. Traditional leaching techniques often rely on strong oxidizing agents, which can be wasteful and environmentally harmful. While cyclic oxidation systems like the $Fe^{3+}/Fe^{2+}$ pair are known, they often fail to achieve high leaching rates, especially when the raw material contains multiple sulfide minerals. In this study, we developed a novel oxidation system using manganese dioxide ($MnO_2$) as the primary oxidizing agent and potassium iodide (KI) as a supporting material to create an $I_2/I^-$ oxidation cycle in a sulfuric acid medium, at an atmospheric pressure between 40 °C and 80 °C. Leaching experiments were conducted under varying temperatures and KI doses. The results demonstrated that for the $MnO_2$-KI system, a zinc leaching degree of 89.78% was achieved after 3 h of leaching at 80 °C, and kinetic studies indicated that the leaching process is diffusion-controlled (through the thin film), with an activation energy of 27.65 kJ mol$^{-1}$. Moreover, this system offers an improved method for separating iodine from the leachate upon completion, enhancing the overall process efficiency. It also opens opportunities to test other primary oxidizing agents in combination with iodide salts. These findings suggest that the $MnO_2$-KI oxidation system offers a promising approach for improving zinc recovery from sphalerite concentrates.

**Keywords:** $I_2/I^-$ cycle; polymetallic materials; zinc extraction; sulfide oxidation; sphalerite

## 1. Introduction

Zinc is a critical industrial metal widely used in galvanization, alloy production, die-casting, and the manufacture of various metal products due to its excellent corrosion resistance and mechanical properties. Efficient extraction of zinc from sphalerite (ZnS), the primary zinc-bearing mineral, is essential to meet global demand and support technological advancement [1]. Sphalerite can sometimes occur in a dispersed form alongside other sulfide minerals and tailings, exhibiting a complex mineralogical composition and a fine-grained structure. It may appear as impregnations, inclusions, and both simple and complex intergrowths within the ore matrix [2]. Traditional hydrometallurgical processes involve leaching sphalerite concentrates in an acidic system, most commonly involving sulfuric acid coupled with oxidizing agents, in order to leach zinc for subsequent recovery. These methods can lead to higher metal recoveries and reduce the hazards associated with air pollution. Key outcomes include the generation of sulfuric acid and/or elemental sulfur and the production of soluble sulfate salts. However, sphalerite's inertness in acid solutions necessitates the use of oxidizing agents to facilitate its oxidative alteration.

Several oxidizing agents have been employed to enhance zinc leaching in acidic media. Hydrogen peroxide ($H_2O_2$) is a strong oxidizer that can significantly accelerate zinc dissolution in sulfuric acid by oxidizing sulfide ions primary to elemental sulfur [3]. Various studies have demonstrated that leaching chalcopyrite and sphalerite with hydrogen peroxide and sulfuric acid can be significantly enhanced by preventing the formation of a compact sulfur layer on the mineral surfaces. This improvement is achieved through the addition of organic compounds like ethylene glycol [4], tetrachloroethylene [5], or liquid paraffin [6], and by mechanical activation [7], which shifts the leaching mechanism to a surface-controlled reaction, resulting in higher metal extractions and increased leaching rates. Iron (III) inorganic salts ($FeCl_3$ [8,9], $Fe_2(SO_4)_3$ [10,11], $Fe(NO_3)_3$ [12,13]) can perform oxidation by converting sulfide ions to elemental sulfur, while iron (III) ions are reduced to iron (II). The usage of these oxidizing agents is well-studied due to the high abundance of the $Fe^{3+}/Fe^{2+}$ redox pair in oxidizing sulfide minerals, both in environmental and artificial circumstances. The iron (III) ions can be regenerated back from iron (II) in the presence of a strong oxidizing agent or metabolic processes induced by iron-oxidizing bacteria [14], creating a cyclic process. Persulfates serve as potent oxidizing agents because they decompose to generate sulfate radicals ($SO_4^{-}$, which are highly reactive species. These sulfate radicals are capable of oxidizing sulfide minerals, thus enabling metal leaching [15,16]. Oxygen is the most abundant oxidizing agent in the atmosphere; unfortunately, it is unable to rapidly oxidize sulfide minerals under atmospheric conditions. However, numerous leaching studies have been performed on elevated pressures and high temperatures in an autoclave in order to test oxygen's potential to perform oxidation in such conditions [17–19].

Nitric acid ($HNO_3$), as well as nitrate salts, serve both as a leaching agent (nitrates are the most numerous water-soluble salts) and an oxidizing agent (thanks to the presence of nitrate anion). However, the use of nitric acid presents challenges, including the formation of nitrogen oxides, which are harmful emissions requiring strict environmental controls [20]. While these oxidizing agents improve zinc leaching rates, they often come with drawbacks such as high reagent consumption, environmental hazards [21] due to the formation of undesirable by-products [22], and increased operational costs [23]. Cyclic oxidation systems like the $Fe^{3+}/Fe^{2+}$ redox pair have been explored to enhance leaching efficiency and leaching rate as well [2,24]. Additionally, other redox pairs discussed in the literature—such as $Cu^{2+}/Cu^{+}$ [25] and the irreversible $Mn^{4+}/Mn^{2+}$ [26] couple—have been investigated primarily in the context of chalcopyrite and its concentrates. Nevertheless, these studies demonstrate the significant influence of these redox pairs on both the leaching degree and rate. The presence of silver within the pyrite crystal lattice has been shown to significantly influence the leaching rate, as silver facilitates the oxidation of sulfides into disulfides via the $Ag^{+}/Ag_2S$ redox couple [27]. However, many ores and concentrates do not contain enough silver to produce this effect, and adding silver salts during the leaching process would be economically unfavorable. All of these systems possess their own limitations in wider application since the overall effect highly depends on initial concentrate composition. Also, these systems may not achieve high leaching rates when processing polymetallic concentrates containing multiple sulfide minerals [28]. The presence of sulfides or polysulfides in the boundary layer can lead to the formation of insoluble elemental sulfur and the precipitation of jarosite if the local pH increases during leaching [22,23]. These occurrences interfere with the oxidation process by forming passivation layers on mineral surfaces, which hinder metal leaching. This limitation underscores the need for alternative oxidation systems capable of overcoming these challenges and improving the efficiency of zinc extraction from complex ores.

In response to these challenges, we propose a novel oxidation system using manganese dioxide ($MnO_2$) as the primary oxidizing agent and potassium iodide (KI) as an auxiliary material to establish an iodine/iodide ($I_2/I^-$) redox cycle in a sulfuric acid medium. $MnO_2$ is a potent oxidizer capable of oxidizing iodide ions to iodine. The generated iodine acts as an effective oxidizing agent for sulfide minerals, oxidizing them and regenerating iodide ions, thus completing the cycle. This $MnO_2$-KI system aims to enhance zinc leaching rates by leveraging the high oxidative potential of iodine while potentially reducing the consumption of strong oxidants and minimizing environmental impact. Moreover, the system offers an improved method for separating iodine from the leachate after leaching, allowing for its recovery and reuse, which enhances the sustainability of the process.

The specific objectives of this research are as follows:

1. Assess the effectiveness of the $MnO_2$-KI system in enhancing zinc extraction compared to traditional oxidizing agents used in acid leaching processes.
2. Investigate the impact of operational parameters, such as temperature and KI dose, on the leaching efficiency and kinetics.
3. Determine the kinetic control mechanism of the leaching process and calculate the activation energy to elucidate the underlying reactions.

By achieving these objectives, this study aims to contribute to the development of more efficient and environmentally friendly hydrometallurgical processes for zinc recovery from sphalerite concentrates. The $MnO_2$-KI oxidation system presents a promising alternative that could overcome the limitations of existing methods, particularly in handling polymetallic sulfide ores with complex compositions. Obtained findings may have broader implications for the extraction of other valuable metals and the advancement of sustainable metallurgical technologies.

## 2. Materials and Methods

The Zn concentrate (ZnS) was collected from a polymetallic ore deposit "Rudnik" flotation plant (Rudnik, Serbia). Prior to characterization and leaching tests, the concentrate was ground by a vibro mill (KHD Humboldt Wedag AG, Cologne, Germany), while all granulometric classes were used in all further experiments.

### *2.1. Structural Characterization*

Surface morphology was determined by scanning electron microscopy (SEM) using a JEOL JSM-7001F (JEOL Ltd., Tokyo, Japan) instrument coupled with an energy-dispersive spectrometer (EDS) (Oxford Xplore 15, High Wycombe, UK). The SEM operated in high vacuum mode (0.1 mPa) at an accelerating voltage of 20 kV and a probe current of 10 nA. The analyzed samples were coated with carbon (15 nm thick layer, density 2.25 g/cm$^3$). For quantitative analyses, the following synthetic standards were used: $FeS_2$ (FeK$\alpha$, SK$\alpha$), ZnS (Zn K$\alpha$, S K$\alpha$), Ni (Ni K$\alpha$), Co (Co K$\alpha$), Cu (Cu K$\alpha$), InSb (Sb K$\alpha$), $Ag_2Te$ (AgL$\alpha$), CdS (Cd L$\alpha$), HgS (Hg M$\alpha$), PbS (Pb M$\alpha$), Bi (Bi M$\alpha$), $CuFeS_2$ (Cu K$\alpha$, Fe K$\alpha$), ZnS (Zn K$\alpha$),Mn (Mn K$\alpha$), Gadolinium Gallium Garnet (Ga K$\alpha$), Ge (Ge L$\alpha$), InAs (As K$\alpha$, In L$\alpha$), $SnO_2$ (Sn L$\alpha$), PbS (Pb M$\alpha$), and Bi (Bi M$\alpha$). EDS detection limit of 2$\sigma$ ~0.3 wt.%.

Additional insights into the structural properties were obtained through qualitative ore microscopic analyses using a polarizing microscope (Carl Zeiss-Jena Axioscope 5 Pol, Jena, Germany) under reflected light to identify metallic and non-metallic minerals. The microscope was equipped with a color camera (Axiocam 105 color, Hebron, KY, USA). Photomicrographs were collected using the software package Carl Zeiss AxioVision SE64 Rel. 4.9.1 (Jena, Germany) with the "Multiphase" module. Objective magnifications of 10×, 20×, 50× (air), and 100× (oil immersion with cedar oil) were used. Photomicrographs accompany the textual descriptions to illustrate the observed textures.

The mineral compositions of the concentrate and leaching residues were determined by X-ray diffraction (XRD) to identify the phase composition. XRD patterns were obtained using a Philips PW 1710/1820 (Eindhoven, The Netherlands) automated diffractometer with a Cu tube operated at 40 kV and 30 mA. The instrument was equipped with a diffracted beam curved graphite monochromator and a xenon-filled proportional counter. Diffraction data were collected over the range of 4–65° 2θ, counting for 1 s per 0.02° step. The divergence and receiving slits were fixed at 1 mm and 0.1 mm, respectively. All XRD measurements were performed at room temperature using a stationary sample holder.

### 2.2. Leaching Experiments

Leaching experiments were performed in a sealed glass reactor (Zhengzhou Great Wall, Zhengzhou, China) equipped with a mechanical stirrer and temperature controller. In a 1.2 L batch reactor (Figure 1), at selected temperature, were placed and mixed initial concentrate (10 g) with 1.5 M analytically prepared water solution of sulfuric acid (Zorka. p. a, Šabac, Serbia). During experiments, a consistent stirring rate at 300 rpm was used. Aliquots were sampled at the desired time and then analyzed after dilution by atomic absorption spectrophotometry (AAS, PERKIN ELMER 703, Shelton, CT, USA). Calibration of the atomic absorption spectrometer was performed using standardized zinc reference solution (Certified Reference Material from CPA Chem, Bogomilovo, Bulgaria) diluted with deionized water. The instrument operated with an acetylene–oxygen flame and a zinc hollow-cathode lamp as the radiation source. All leaching experiments were carried out in triplicate to minimize measurement uncertainty, and the reported zinc concentrations represent the average of the three measurements. The AAS measurement served as the basis for calculating the zinc extraction rates (%). In some experimental setups, auxiliary oxidation agents—manganese dioxide (Alkaloid, purity > 95%, Skopje, North Macedonia) and potassium iodide (Darmstadt, Merck, purity > 95%, Germany) were added. During the leaching tests, different operational parameters were varied, such as time (0–180 min), temperature (40, 60, 70, and 80 °C) and presence of potassium iodide (1, 2, and 3 wt.% of the initial ZnS mass).

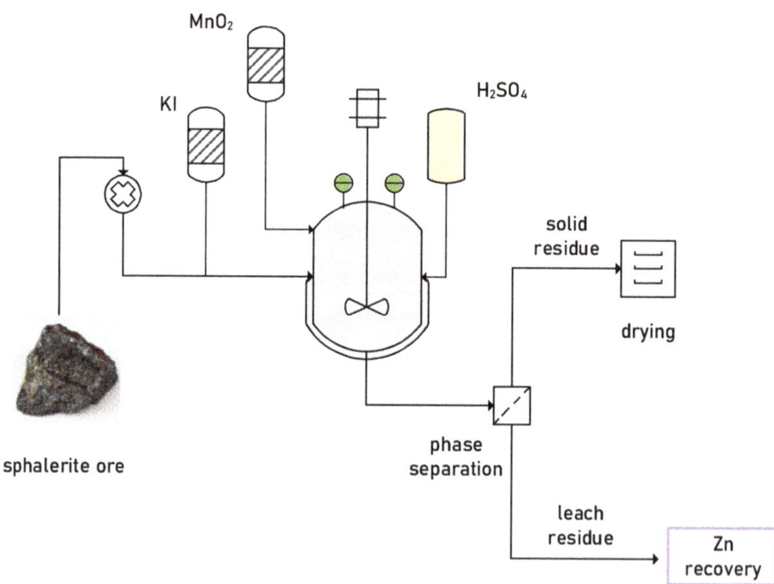


**Figure 1.** Simplified scheme of leaching procedure.

The chemical composition of the ZnS sample (Table 1) was ascertained by acid digestion followed by AAS.

**Table 1.** Chemical composition of Zn concentrate.

| Element | Zn | Cu | Fe | As | Cd | Pb | S |
|---|---|---|---|---|---|---|---|
| (wt.%) | 46.91 | 0.806 | 6.77 | $1.8 \times 10^{-4}$ | 0.338 | 0.287 | 27.87 |

After completing each experiment, the solution was filtered to separate the leaching residues. The solid residues were collected, rinsed twice with distilled water, and dried in an oven. The prepared samples were then thoroughly analyzed.

Better insights into observed chemical reactions can give kinetic models. In this work, we employed several equations to linearize the experimental findings using the modified Sharp's method of reduced half reaction time [29]. Principles are detailed and described in our previous publications [30–32].

Chemical processes' activation energy (Ea), which is typically determined using the Arrhenius equation, frequently offers significant knowledge into the mechanisms underlying the reactions:

$$k = A * \exp(-Ea/RT) \quad (1)$$

where T is the absolute temperature (K), R is the gas constant (8.314 $J \cdot mol^{-1} \cdot K^{-1}$), Ea is the activation energy ($J \cdot mol^{-1}$), A is the pre-exponential constant, and k is the rate constant ($s^{-1}$). Plotting ln(k) against 1/T yields the slope of the linear fit, from which the activation energy may be calculated.

## 3. Results and Discussion

### *3.1. Material Characterization*

The sphalerite concentrate samples used in this study, as well as the samples obtained after leaching, were fully characterized to gain insight into the initial mineralogical and morphological structures. Characterization of the leaching residues, on the other hand, is very helpful in determining the leaching mechanism and is often crucial for further optimization of the leaching process. Optical and electron microscopy provided valuable insights into both the morphological structure and mineralogical composition of the ZnS sample and its solid residues; mineralogical composition was also determined using XRD. These analytical techniques allowed us to observe surface features and textural relationships within the samples.

#### 3.1.1. Optical Microscopy

Under the optical microscope (Figure 2), it was observed that approximately 80% of the sphalerite occurs as free grains, with the remaining portion present in simple to complex intergrowths with other minerals. Chalcopyrite is most frequently associated with sphalerite in this concentrate, followed by pyrite, galena, gangue minerals, and chalcopyrite impregnations. The grain sizes of sphalerite vary from roughly 10 μm to over 350 μm. In addition to these minerals, particles of gold and silver were also detected.

Chalcopyrite is the next most abundant mineral in the concentrate. It occurs as simple to complex assemblages with sphalerite, pyrite, galena, and gangue minerals, or as oval, micronic impregnations within sphalerite grains. Additionally, chalcopyrite may appear as separate grains, accounting for about 65–70% of its occurrence. No discernible free gold or silver was found associated with chalcopyrite.

After chalcopyrite, pyrite is the most prevalent mineral (Supplementary Materials, Figure S1). The grain size varies between approximately 10 μm and 150 μm. It can be found in simple to complex assemblages with sphalerite, chalcopyrite, and tailings minerals, or it can be found in independent grains (75%). There is no discernible self-propelled silver or gold in pyrite. The free pyrite grain is roughly 150 μm in size.

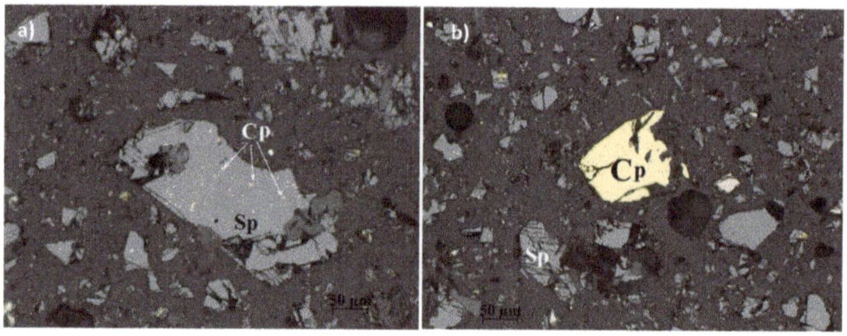


**Figure 2.** Sphalerite (Sp) grain with "chalcopyrite disease" (Cp) (**a**); Free chalcopyrite and sphalerite grains (**b**).

The amount of gold or silver in traces (Figure 3) was found in the material sample under optical analysis. It can be found in sphalerite and tailings minerals, in which can be as large as 6–7 μm.

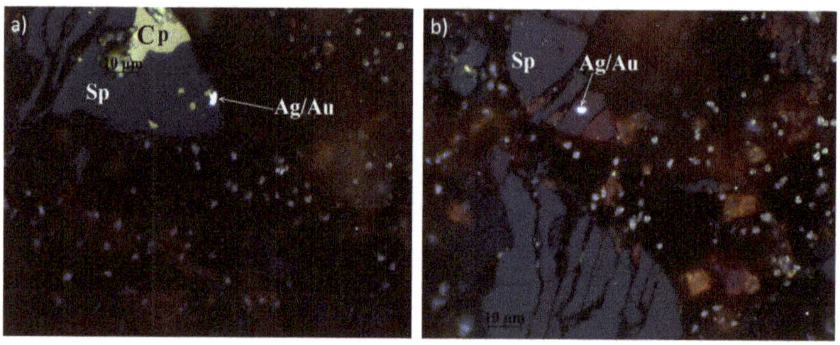


**Figure 3.** Native electrum (Ag/Au) in Sp (**a**); Native silver/gold in addition to sphalerite in the ZnS sample (**b**).

### 3.1.2. XRD Analysis

The polycrystalline sample (powder) was examined using the X-ray diffraction method. The following minerals were discovered to be present in the examined sample: chalcopyrite, pyrite, quartz, and sphalerite (Figure 4). The most prevalent mineral was undoubtedly sphalerite, whereas chalcopyrite, pyrite, and quartz were far less common.

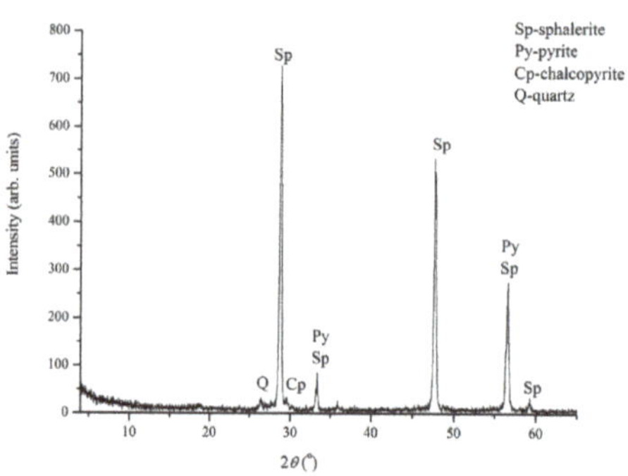


**Figure 4.** Powder diffractogram of the ZnS sample.

3.1.3. Field Emission SEM

A SEM image (Figure 5) shows that concentrate particles have sharp, smooth edges. Constituents are in the diameter range 50–100 μm. The presence of minerals like sphalerite, pyrite, and galena were confirmed via SEM-EDS analysis as well as chemical and XRD analysis.

**Figure 5.** SEM micrograph of ZnS sample.

## 3.2. Leaching Results

### 3.2.1. Leaching with $MnO_2$

Initial tests were conducted using only the primary oxidizing agent, manganese dioxide ($MnO_2$), in a sulfuric acid solution under controlled temperature conditions (80 °C). Even at the highest temperature tested, the extraction yield was only 41.95%, indicating that the extraction process was not sufficiently effective. This low yield occurred despite the sufficient amount of $MnO_2$ (1:2) and efficient mixing (300 rpm), suggesting that factors other than reagent concentration and mixing were limiting the extraction.

One possible explanation for the low extraction efficiency is the hindrance of electron transfer between the oxidizing agent and the sulfur species generated from the sulfide minerals. This impediment could prevent the effective oxidation of sulfide minerals, thereby reducing the extraction yield.

Evidence supporting this explanation was obtained through SEM-EDS analysis (Figure 6). The spectrum obtained from the surfaces of marcasite grains revealed a non-stoichiometric sulfur content—specifically, the sulfur content was lower than expected for the chemical composition of marcasite ($FeS_2$). This suggests that disulfide anions ($S_2^{2-}$) in pyrite are partially oxidized, leading to the formation of marcasite and elemental sulfur. The elemental sulfur forms a film that can subsequently continue to oxidize to higher oxidation states. Thanks to the electro nonconductive layer of sulfur, further oxidation of sphalerite is limited by the diffusion of both zinc and sulfide into the bulk solution and further oxidation to thiosulfate and sulfite ion species.

Moreover, SEM-EDS analysis of the pyrolusite ($MnO_2$) grains (Figure 6d,e) showed that their chemical composition did not match that of pure $MnO_2$. The oxygen content was less than expected, indicating that manganese on the surface of the grains had been reduced to lower oxidation states ($Mn^{3+}$ and $Mn^{2+}$). This reduction of $MnO_2$ suggests that it participated in redox reactions; however, the overall extraction yield of zinc remained low, all due to ineffective electron transfer from sphalerite to pyrolusite surface.

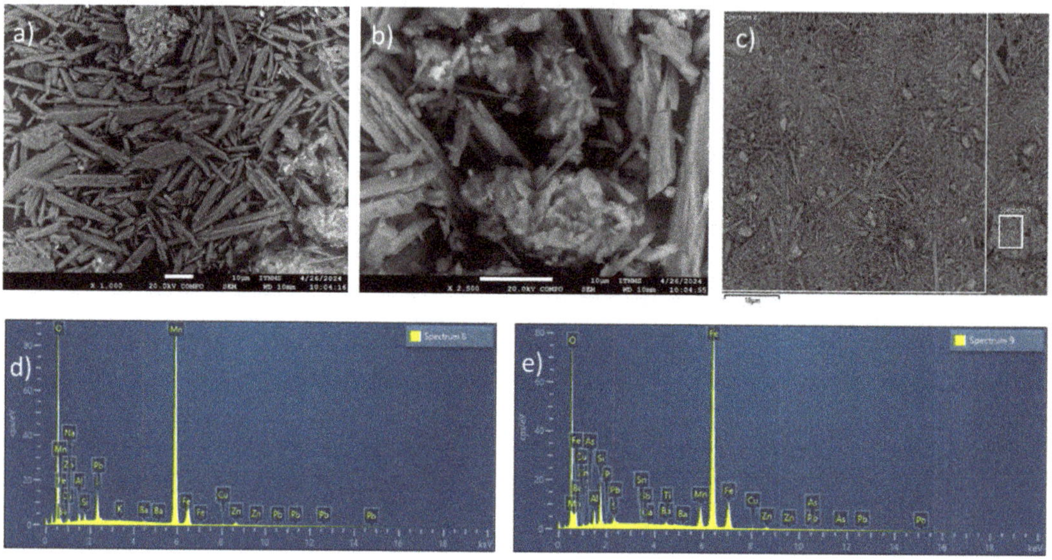

Figure 6. SEM and SEM-EDS spectra of leaching residues in system ZnS + $MnO_2$ + $H_2SO_4$: (a) 1000× (b) 2500×; (c) microfigure with selected areas for EDS analysis; (d) EDS results of spectrum 8; (e) EDS results of spectrum 9.

In Figure 6 are presented SEM and SEM-EDS spectra of the leaching residue sample, illustrating these findings.

The chemical reaction that occurs during leaching in the presence of the primary oxidizing agent alone is as follows (Equation (2)):

$$ZnS + 4MnO_2 + 4H_2SO_4 = ZnSO_4 + 4MnSO_4 + 4H_2O \quad (2)$$

Scans show that added $MnO_2$ was not fully reacted, which is expected since a primary oxidizing agent is inserted in sufficient amounts, and the leaching degree did not reach maximum value. Therefore, as described previously, manganese dioxide was unable to achieve complete oxidation of the present sulfide minerals.

3.2.2. Leaching in System with $MnO_2$ and KI

Impact of Temperature on Leaching Degree

The efficiency of leaching zinc (Zn) from ZnS increases with temperature, from 40 °C to 80 °C, a trend consistent with the principles of chemical kinetics. Higher temperatures supply additional thermal energy, which enhances the movement of particles and the frequency of effective collisions between sulfide minerals and intermediary oxidant ($I_2$). This leads to accelerated reaction rates and, consequently, a more efficient leaching of zinc into the leaching medium.

An intriguing observation from the XRD analysis is the absence of crystalline sulfur when the leaching is performed at lower temperatures (Figure 7).

The qualitative mineralogical composition of the analyzed sample (Figure 7) is as follows: pyrolusite, sphalerite, chalcopyrite, marcasite, and quartz. The most abundant phase is pyrolusite; sphalerite is less represented, while quartz, chalcopyrite, and marcasite are the least represented. All other crystal phases are below detection limits. Quantitative mineralogical analysis is as follows: pyrolusite 54.3 wt.%, sphalerite 41.2 wt.%, chalcopyrite 1.1 wt.%, marcasite 1.0 wt.%, quartz 0.8 wt.%, sulfur 0.6 wt.%, calcite 0.4 wt.%, galena 0.3 wt.%, and barite 0.3 wt.%. These results were obtained in accordance with mineralogical and chemical analysis.

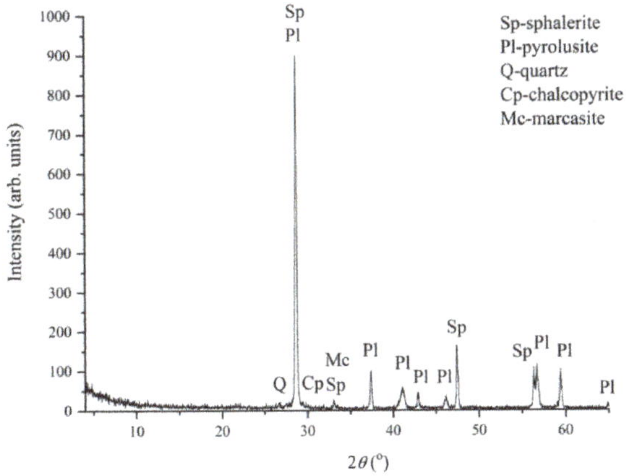


**Figure 7.** XRD pattern with legend for leaching residue; experimental parameters (3 wt.% KI + stoichiometric amount of $MnO_2$ at 40 °C).

However, the traces of sulfur grains are detected in reflected light microscopy images (Figure 8a). From the other reflected light microscopy images (Figure 8), it can be observed that the most abundant phase is pyrolusite; sphalerite occurs less often, while chalcopyrite, marcasite, and quartz are the least represented. Galena is present in traces.

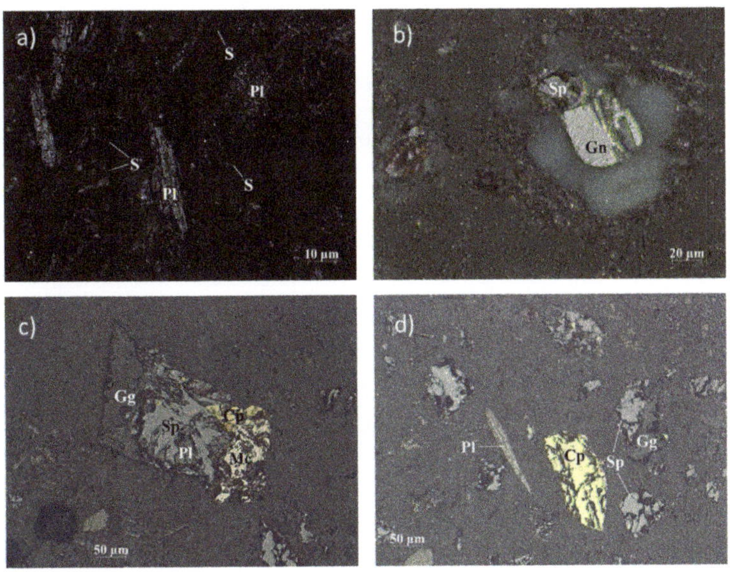


**Figure 8.** Reflected light microscopy image of the sample at 40 °C: Liberated grains of pyrolusite (Pl) and small sulfur traces (**a**); Intergrowth of sphalerite (Sp) and galena (Gn) (**b**); Intergrowth of sphalerite (Sp), chalcopyrite (Cp), marcasite (Mc), pyrolusite (Pl) and gangue (Gg) (**c**); Intergrowth of sphalerite (Sp) and gangue (Gg), "chalcopyrite (Cp) disease" in sphalerite, and liberated pyrolusite (Pl) crystal (**d**).

Pyrolusite occurs mainly in the form of lath-like habitus up to 200 μm in length and, somewhat less often in irregular grains up to 100 μm in length. Sphalerite is present as irregular liberated grains (significantly transformed and eroded grains), up to 150 μm in length. It often contains "chalcopyrite disease" and inclusions of chalcopyrite up to 50 μm in size. Intergrowths of sphalerite with chalcopyrite, marcasite, pyrolusite, or galena occur to a lesser extent. Very rarely (only in one place), a complex intergrowth of sphalerite, chalcopyrite, marcasite, pyrolusite, and quartz, up to 300 μm in length, as well as a complex intergrowth of sphalerite, chalcopyrite, and galena are present. Except in the mentioned way, chalcopyrite rarely occurs in the form of liberated irregular grains up to 150 μm in length. Marcasite is present in the form of liberated grains up to 150 μm in length. Galena occurs in irregular grains up to 60 μm in length.

One noteworthy observation is an accidentally identified presence of galena (PbS) surrounded by anglesite, as shown in Figure 8. Most likely, anglesite is formed due to the oxidation of sulfide ions and consequential liberation of lead ions and precipitation with readily present sulfate anions. The presence of galena in the sphalerite concentrate is expected since the initial ore contains both sphalerite and galena that in some cases are very difficult to separate thanks to intergrowths of sphalerite and galena, especially if the grains are very small (under 80 μm). Therefore, identifying galena as well as anglesite in the leaching residue is not surprising.

At a higher temperature, 80 °C (Figure 9), the XRD method detected the presence of crystalline sulfur phases. This phenomenon can be explained by the temperature-dependent kinetics of sulfur formation and crystallization during the leaching process. At higher temperatures, the oxidation reactions not only proceed more rapidly but also enable the possibility of elemental sulfur crystallization. The increased thermal energy enables sulfur atoms to rapidly migrate from the sphalerite particle surface where, thanks to their high amount and increased temperature, they form nucleation centers and grow undisturbedly into crystalline structures large enough to be detected by XRD.

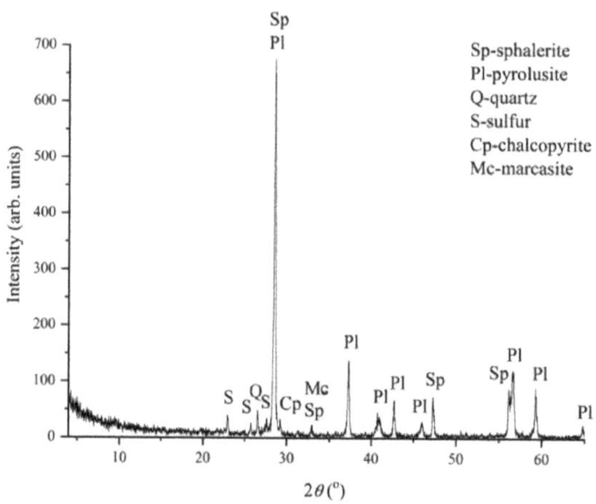


**Figure 9.** XRD pattern with legend for leaching residue; experimental parameters (3 wt.% KI + stoichiometric amount of $MnO_2$ at 80 °C).

The qualitative mineralogical composition of the analyzed sample (Figure 9) is as follows: pyrolusite, sphalerite, quartz, chalcopyrite, marcasite, and sulfur. The most abundant phase is pyrolusite; sphalerite is less represented, while the least abundant are quartz, sulfur, chalcopyrite, and marcasite. All other crystal phases are below detection limits.

Quantitative mineralogical analysis is as follows (weight): pyrolusite 57 wt.%, sphalerite 38 wt.%, chalcopyrite 1.1 wt.%, sulfur 1.0 wt.%, marcasite 1.0 wt.%, quartz 0.9 wt.%, calcite 0.4 wt.%, galena 0.3 wt.%, and barite 0.3 wt.%. These results were obtained via software (Carl Zeiss AxioVision SE64 Rel, Jena, Germany. 4.9.1 with the "Multiphase" module) in accordance with chemical analysis.

A SEM image of leaching residue obtained after leaching at 80 °C is given in Figure 10, while the EDS spectra are given in the Supplementary Materials, Figures S3 and S4. Calculations based on obtained EDS spectra (Figures S3 and S4) are given in Table 2.

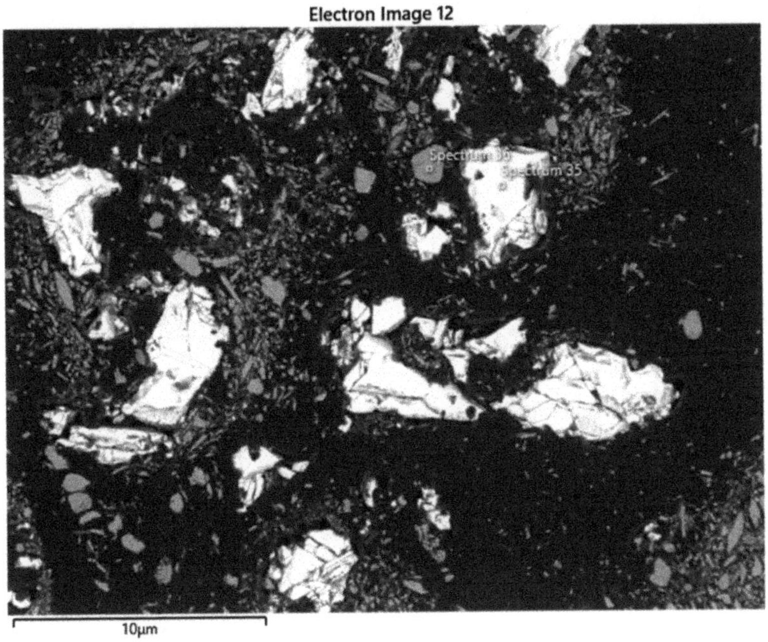


**Figure 10.** SEM photomicrograph of the leaching residue sample at 80 °C.

**Table 2.** EDS calculation for obtained SEM (Figure 10) Spectrum 35 and 36.

| Element | Spectrum 35 (Sphalerite) | | Spectrum 36 (Pyrolusite) | |
|---|---|---|---|---|
| | Weight% | Atomic% | Weight% | Atomic% |
| O | - | - | 29.90 | 59.43 |
| S | 33.58 | 50.05 | - | - |
| Mn | 0.80 | 0.69 | 70.10 | 40.57 |
| Fe | 11.41 | 9.77 | - | - |
| Zn | 53.73 | 39.29 | - | - |
| Cd | 0.48 | 0.20 | - | - |
| Total | 100.00 | 100.00 | 100.00 | 100.00 |

In contrast, at lower temperature, the elemental sulfur produced may remain in an amorphous or finely dispersed state. The lower temperature (40 °C) is insufficient to promote significant crystallization of sulfur. Amorphous sulfur lacks the long-range atomic order required for detection by XRD (Figure 7), and very fine sulfur particles may fall below the detection limit of the instrument. Therefore, despite the production of sulfur during the leaching process at lower temperatures, it may not be observable in XRD patterns due to its amorphous nature or small particle size. These findings highlight the complex interplay between temperature, reaction kinetics, and phase formation. While higher temperatures enhance Zn leaching efficiency by accelerating reaction rates, they also promote the crystallization of sulfur, which could potentially impact the leaching process by forming passivation layers. However, the presence of sulfur was confirmed even on the leaching residue obtained after leaching at 40 °C but only when reflected light and scanning electron microscopy methods were employed. In Figure 11a can be seen a small detail at the bottom left that comes from elemental sulfur.

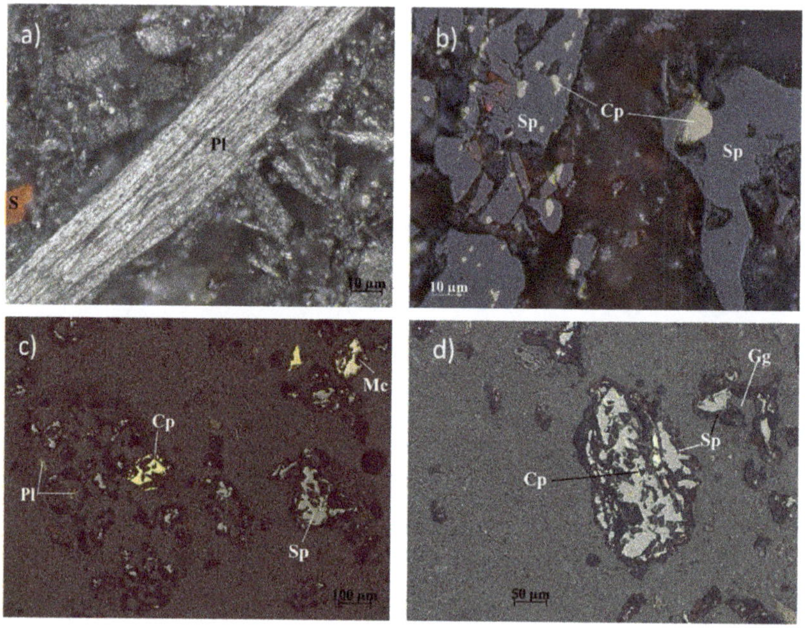


**Figure 11.** Reflected light microscopy image of sample at 80 °C: Liberated pyrolusite crystal grains (Pl), sulfur (S) (**a**); "Chalcopyrite disease" (Cp) and inclusion in sphalerite (Sp) (**b**); Liberated grains of sphalerite (Sp), chalcopyrite (Cp), marcasite (Mc), and pyrolusite (Pl) (**c**); Intergrowth of sphalerite (Sp) and gangue (Gg), "chalcopyrite disease" (Cp) in sphalerite (**d**).

Reflected light microscopy images (Figure 11) revealed that the most abundant phase is pyrolusite; sphalerite occurs less often, while chalcopyrite, marcasite, sulfur, and quartz are the least represented. Galena is present in traces. Pyrolusite occurs mainly in the form of lath-like habitus up to 200 µm in length and somewhat less often in irregular grains up to 100 µm in length. Sphalerite is present as irregular liberated grains (significantly transformed and eroded grains), up to 250 µm in length. It often contains "chalcopyrite disease" and inclusions of chalcopyrite up to 50 µm in length. Intergrowth of sphalerite and chalcopyrite occurs to a lesser extent. Except in the mentioned way, chalcopyrite rarely occurs as liberated irregular grains up to 100 µm in length. Marcasite is present in the form of liberated grains up to 150 µm in length. Sulfur occurs in the form of irregular grains. Sphalerite contains Cd up to 0.5 wt.%. Gangue minerals are represented by quartz, and to a lesser extent by feldspar, barite, and calcite.

SEM image of the leaching residue obtained after leaching at 40 °C is given in Figure 12, while the EDS spectra are given in Supplementary Materials, Figures S5–S8. Calculations based on obtained EDS spectra (Figures S5–S8) are given in Table 3.

Electron Image 19

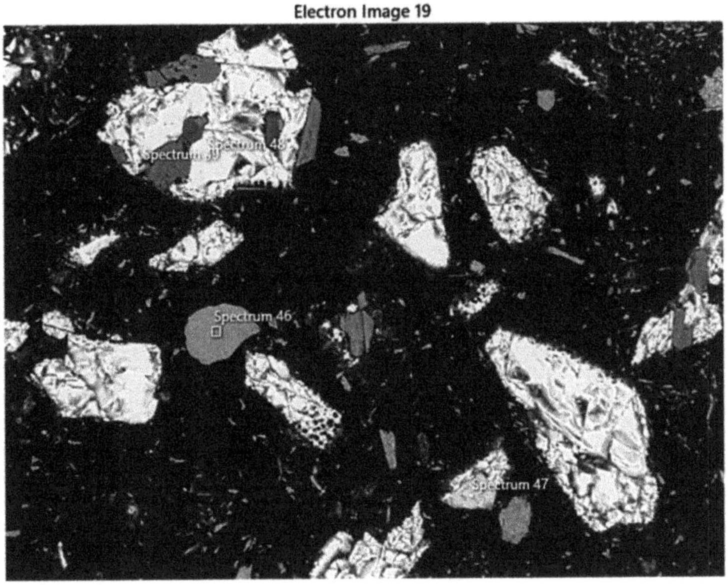


**Figure 12.** SEM photomicrograph of the leaching residue sample at 40 °C.

**Table 3.** EDS calculation for obtained SEM (Figure 12) Spectrum 46–49.

| Element | Spectrum 46 (Pyrolusite) | | Spectrum 47 (Marcasite) | | Spectrum 48 (Sphalerite) | | Spectrum 49 (Chalcopyrite) | |
|---|---|---|---|---|---|---|---|---|
| | Weight% | Atomic% | Weight% | Weight% | Weight% | Atomic% | Weight% | Atomic% |
| O | 30.55 | 60.11 | - | - | - | - | - | - |
| S | 0.23 | 0.23 | 39.86 | 53.58 | 33.42 | 49.88 | 34.79 | 49.86 |
| Mn | 69.21 | 39.66 | 0.50 | 0.39 | 0.68 | 0.60 | 0.67 | 0.56 |
| Fe | - | - | 59.65 | 46.03 | 11.26 | 9.65 | 29.46 | 24.24 |
| Cu | - | - | - | - | - | - | 33.11 | 23.95 |
| Zn | - | - | - | - | 54.22 | 39.69 | 1.97 | 1.38 |
| Cd | - | - | - | - | 0.42 | 0.18 | - | - |
| Total | 100.00 | 100.00 | 100.00 | 100.00 | 100.00 | 100.00 | 100.00 | 100.00 |

Figure 13 illustrates the degree of zinc (Zn) leaching $\alpha$ (%) from the concentrate over time. The chemical reaction for sphalerite (ZnS) oxidation is the same as in the case of leaching zinc in the presence of only manganese dioxide ($MnO_2$) (Equation (2)).

It has been shown that increasing the temperature raises the yield of the observed reaction (Equation (2)). However, since the composition of the reaction mixture differs from the previous case, an additional chemical reaction occurs (Equation (3)):

$$MnO_2 + 2KI + 2H_2SO_4 = MnSO_4 + K_2SO_4 + I_2 + 2H_2O \quad (3)$$

The iodine ($I_2$) formed during reaction (3) reacts further by capturing a single iodide ion ($I^-$) [33] and participates in another chemical reaction, as reported in the literature [34]. For simplicity, we represent this subsequent reaction by considering only the participation of iodine as a reagent (Equation (4)):

$$ZnS + I_2 = Zn^{2+} + S + 2I^- \quad (4)$$

Equations (3) and (4) alternate, facilitating further oxidation of sphalerite in the leaching medium. However, there are no significant differences between the SEM-EDS images or the optical microscope micrographs obtained after leaching with $MnO_2$ in sulfuric acid alone and those obtained when the primary oxidizing agent is combined with potassium iodide (KI). This observation suggests that the leaching mechanisms in both cases are very similar and that the solid leaching residue products are almost the same.

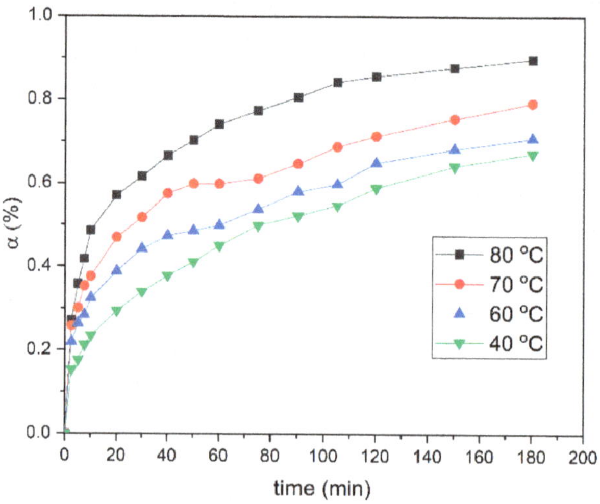


**Figure 13.** Influence of temperature and reaction time on leaching degree of Zn (3 wt.% KI + stoichiometric amount of $MnO_2$).

Impact of KI Concentration on Leaching Efficiency

The efficiency of leaching zinc (Zn) from sphalerite concentrates exhibits a non-linear relationship with the concentration of potassium iodide (KI) added to the leaching medium at 80 °C (Figure 14). This intriguing pattern suggests a complex interplay between the chemical reactions involved and the concentration of KI, which influences the availability and effectiveness of the oxidizing agents in the leaching process.

When the amount of KI is low (1 wt.% to the mass of sphalerite concentrate), the system benefits from an optimal balance where the iodine ($I_2$) produced in reaction (3) is sufficient to enhance the oxidation of sphalerite in reaction (4) without causing adverse side effects.

At this concentration of iodine, the amount of elemental sulfur produced does not significantly hinder the leaching process, resulting in a high Zn leaching rate (79.04%).

When the KI concentration is increased to 2 wt.%, the Zn leaching rate unexpectedly decreases to 67.53%. This reduction can be attributed to the intensified formation of a passivation layer of elemental sulfur on the sphalerite surface. The higher concentration of iodine leads to increased production of elemental sulfur (S), which can accumulate and form a barrier that impedes further access of the oxidizing agents to the mineral surface. This increase in effective oxidant concentration, coupled with the passivation effect, results in lower leaching efficiency.

At the highest amount of KI (3 wt.%), the Zn leaching efficiency markedly increases to the highest determined value of 89.78%. The higher concentration of iodine now becomes sufficient not only to oxidize ZnS but also to react with the elemental sulfur passivation layer, converting it into soluble sulfur compounds with higher oxidation states, such as sulfur dioxide ($SO_{2(aq)}$) and further oxidation of thiosulfate ions to tetrathionate ion that eventually are oxidized to sulfate ion species. The oxidation of elemental sulfur can be represented by the following reaction (5):

$$S_{(s)} + 2H_2O_{(l)} + 2I_{3(aq)}^{-} \rightarrow SO_{2(aq)} + 6I_{(aq)}^{-} + 4H^{+}_{(aq)} \quad (5)$$

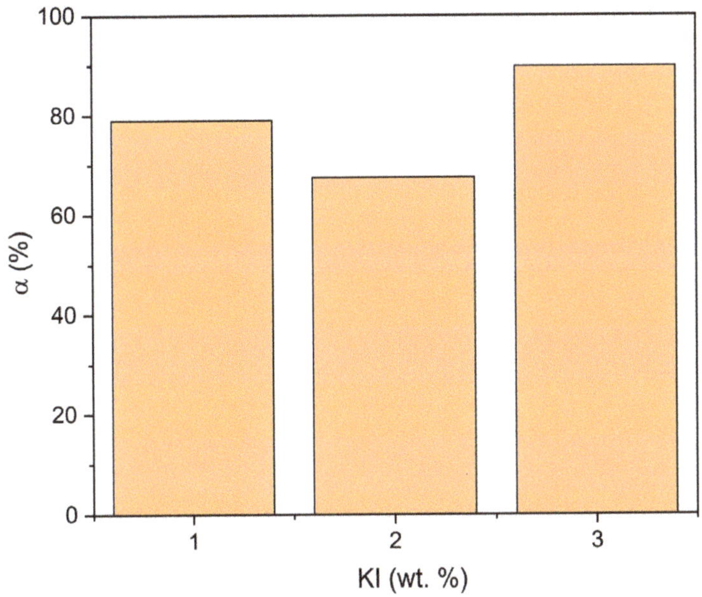


**Figure 14.** Effect of different KI contents on Zn leaching efficiency (temperature interval 40–80 °C).

Sulfur dioxide formed in reaction (5) then follows two oxidation pathways:

- Either it captures readily present sulfur on the sphalerite particle surface leading to the formation of thionate ion species, Equation (6):

$$SO_{2(aq)} + H_2O_{(l)} + S_{(s)} \rightarrow H^{+}_{(aq)} + HS_2O_{3(aq)}^{-} \quad (6)$$

The thionate ions subsequently are oxidized with iodine, leading to the formation of tetrathionate ion species, reaction (7):

$$2\,HS_2O_{3(aq)}^{-} + I_{3(aq)}^{-} \rightarrow 3I_{(aq)}^{-} + HS_4O_{6(aq)}^{-} + H^{+}_{(aq)} \quad (7)$$

Both thiosulfate and tetrathionate ion species are eventually catalytically [35] or directly chemically [36] oxidized to sulfates; both reactions are feasible in the observed system thanks to both the presence of chalcopyrite required for catalytical oxidation and iron (III) ion specimen required for chemical oxidation.

- Also, sulfurous acid ($SO_{2(aq)}$) may be oxidized to the S(V) oxidation state with manganese (IV) oxide, reaction (8) [37]:

$$2MnO_{2(s)} + 3SO_{2(aq)} \rightarrow MnS_2O_{6(aq)} + MnSO_{4(aq)} \quad (8)$$

Manganese dithionate participates in further oxidation to $MnSO_4$ with readily available $MnO_2$ or triiodine ions, reactions (9) [37] and (10), respectively:

$$MnS_2O_{6(aq)} + MnO_{2(s)} \rightarrow 2MnSO_{4(aq)} \quad (9)$$

$$MnS_2O_{6(aq)} + I_{3(aq)}^{-} + 2H_2O_{(l)} \rightarrow MnSO_{4(aq)} + HSO_{4(aq)}^{-} + 3I_{(aq)} + 4H^{+}_{(aq)} \quad (10)$$

A higher initial concentration of potassium iodide leads to the higher concentration of iodine required for oxidation of the passivation layer, and the sphalerite surface is re-exposed to the leaching agents, allowing the leaching process to proceed more efficiently. Furthermore, at this KI initial content, the system's redox potential is enhanced, and the equilibrium may shift back in favor of free iodine thanks to the presence of excess iodide ions. This ensures a sufficient supply of the active oxidant to sustain the leaching reaction, overcoming the limitations observed at 2 wt.% KI.

### 3.3. Leaching Kinetics

To model the leaching kinetics, various kinetic models encompassed within the modified Sharp's method were employed to describe the system's behavior, as depicted in Figure 15. Among these, the Jander equation [29] for three-dimensional diffusion through a product layer (depicted by navy blue triangle symbols on Figure 15) provided the best fit to the experimental data at 80 °C, yielding a coefficient of determination ($R^2$) of 0.98. This high correlation indicates that the Jander model accurately describes the leaching process under the studied conditions. Consequently, this model was selected for fitting all leaching experiment data obtained at various temperatures.

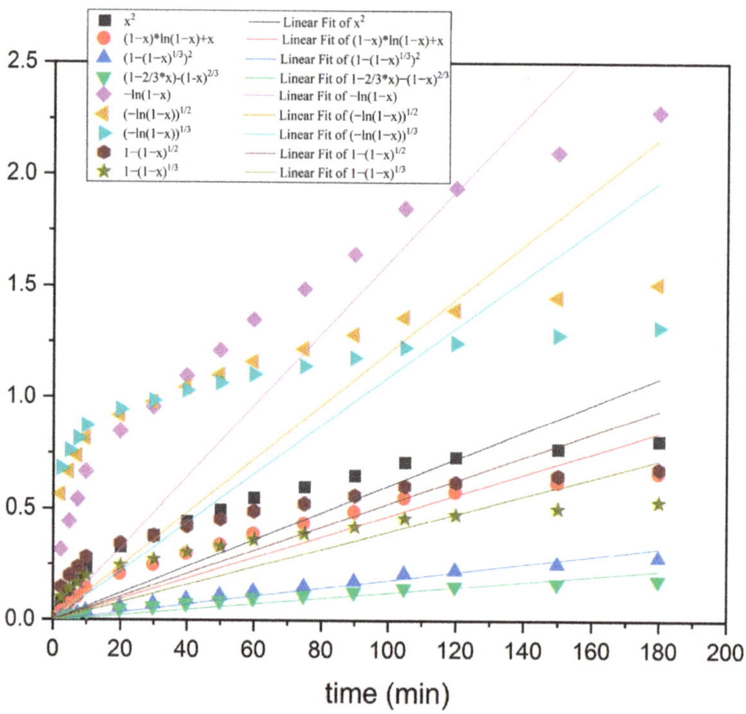


**Figure 15.** Experimental data correlated with modified Sharp's method.

The results suggest that the rate of the leaching reaction is limited by the diffusion of zinc and sulfide species through a product layer—most likely composed of elemental sulfur formed on the surface of sphalerite particles. This product layer acts as a barrier, impeding mass transfer and slowing down the leaching rate [38]. The formation of such a sulfur layer is consistent with findings reported in the literature, where elemental sulfur deposition on sphalerite particles during leaching processes has been observed to significantly decrease the leaching yield of the target metals [39,40].

In Figure 16 are presented the fitted experimental data obtained from leaching ZnS in a system containing $MnO_2$ and 3 wt.% KI over a temperature range of 40 to 80 °C, using the Jander equation. The experimental data across the entire temperature range exhibit a high coefficient of determination with this model ($R^2 > 0.95$), indicating an excellent correlation. The calculated leaching rate constants are $5.35 \times 10^{-4}$ $min^{-1}$ and $1.78 \times 10^{-3}$ $min^{-1}$ at 40 °C and 80 °C, respectively, by applying the Jander equation in the modified Sharp's method. These values demonstrate that increasing the temperature significantly enhances the leaching rate, making it more than three times faster at 80 °C compared to 40 °C (Figure 16).

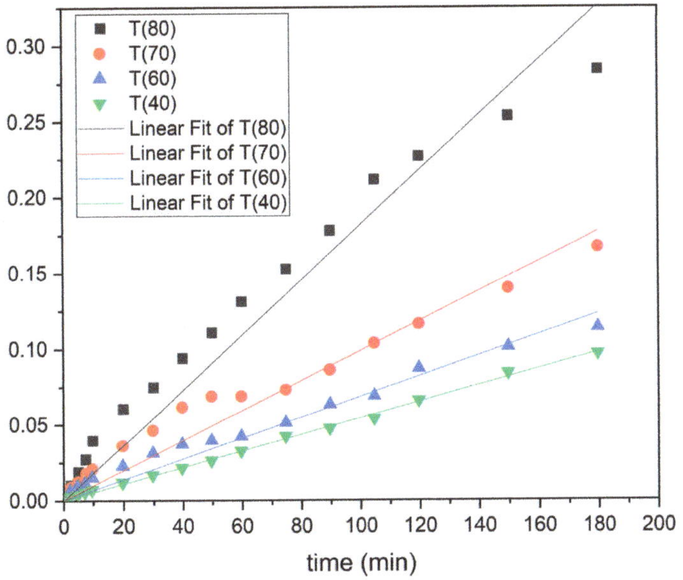


**Figure 16.** Modeling experimental data obtained at different temperatures utilizing the Jander equation.

Since the Jander equation describes three-dimensional diffusion through a product layer, the strong agreement between the model and the experimental data suggests that the rate-limiting step is the diffusion of species through a product layer—most likely a passivation layer of elemental sulfur formed on the surface of sphalerite particles. The formation of this layer occurs at all temperatures, although the amount and its impact on the leaching degree may vary. However, the crystallization of sulfur is confirmed only at the highest leaching temperature (at 80 °C, Figures 9 and 11) but the presence of sulfur traces is confirmed even at the lowest leaching temperature (at 40 °C, Figures 7 and 8).

As previously discussed, higher leaching rates and elevated temperatures facilitate sulfur crystallization, which can exacerbate the passivation effect by forming a more coherent and impermeable layer. This issue can be mitigated by adding a sufficient amount of potassium iodide (>3 wt.%), which increases the concentration of iodine in the leaching medium. The increased iodine concentration enables the oxidation of the formed sulfur passivation layer, as iodine effectively converts elemental sulfur into soluble sulfur compounds. This oxidation diminishes the mass transfer barrier and enhances the leaching efficiency, maintaining high leaching rates even at elevated temperatures. Similar strategies have been reported in the literature, where the addition of oxidizing agents helps to prevent passivation and improve leaching outcomes [41,42].

The final step of the kinetics calculations involved determining the activation energy (Ea), which was obtained using the Arrhenius equation in its linearized form. In Figure 17, the dependence of ln(k/min$^{-1}$) versus 1/T is presented, and the data show a strong linear correlation ($R^2 > 0.89$). The activation energy was calculated from the slope of this linear function, as described by Equation (11):

$$Ea = -R * Slope \quad (11)$$

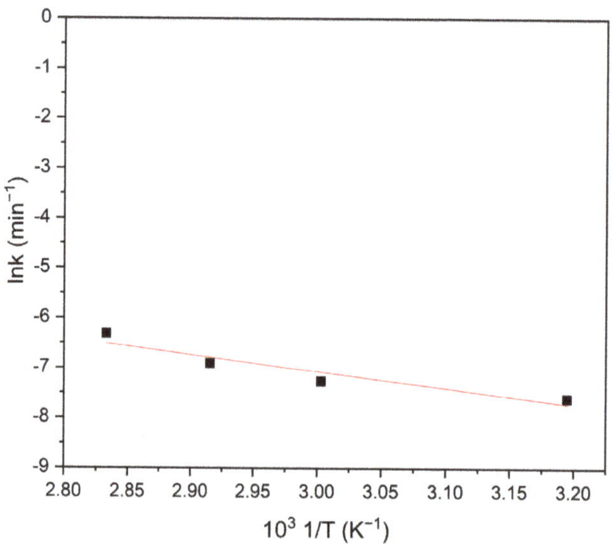


**Figure 17.** Calculated energy of activation (temperature interval 40–80 °C).

The calculated value of 27.65 kJ/mol indicates that diffusion is the limiting step in the leaching reaction, also confirmed elsewhere [39–41].

This relatively low activation energy suggests that the oxidation process proceeds with a low energy barrier, allowing leaching to occur rapidly at moderate temperatures. Moreover, the low value of the activation energy is consistent with the experimental data, where the best fitting was achieved using the Jander equation. Since this model provided the highest goodness of fit, the low activation energy value also confirms that leaching is a diffusion-controlled process. Similar findings have been reported in the literature, where diffusion mechanisms were identified as the controlling step in comparable leaching systems.

Some previous studies that were performed in order to determine the Zn leaching rate and the limitation step that controls the extraction from sphalerite concentrate are given in the overview table (Table 4). Results of some studies given in Table 4 seem more promising than those obtained in our research. For example, in a study of Peng et al., the authors utilized both a nitration mixture and tetrachloroethylene as a leaching medium; the limitations of this study are mostly related to environmental and separation technique concerns. Also, a study performed by Santos et al. achieved a very high leaching rate (95.1%); however, they achieved such high leaching rates after 25 h of leaching, which is energy-consuming for practical applications. The same issue is associated with the study of Babu.

**Table 4.** Literature data of leaching efficiency and mechanism in wide temperature range on sphalerite concentrate samples using various oxidizing agents.

| Solution | Temperature, °C | Leaching Rate, % | Process That Determines the Reaction Rate | Bibliography |
|---|---|---|---|---|
| $H_2SO_4$–$HNO_3$ | 45–85 | 70.1–99.6 | diffusion control | [43] |
| $H_2SO_4$–$Fe_2(SO_4)_3$ | 70–110 | 87.3–89.1 | chemical control | [10] |
| $H_2SO_4$–$(NH_4)_2S_2O_8$ | 30–70 | 44.2–96.1 | diffusion control | [15] |
| $H_2SO_4$–$Fe_2(SO_4)_3$ | 40–80 | 22.06–89.76 | mixed control | [44] |
| | | | diffusion control | |
| $H_2SO_4$–$Fe_2(SO_4)_3$-NaCl | 20–95 | 23.8–95.1 | chemical control | [45] |
| | | | mixed control | |
| $H_2SO_4$–$MnO_2$–KI | 40–80 | 67.53–89.73 | diffusion control | Current study |

Finally, Table 5 provides a rough material cost analysis for three different leaching scenarios applied to the sphalerite concentrate (ZnS sample). In each scenario, a distinct oxidizing agent—ammonium persulfate, iron(III) sulfate, or manganese dioxide with potassium iodide—was used. For simplicity, the costs of sulfuric acid, sodium chloride, and iodine are excluded from all three scenarios, as their prices or required quantities are relatively low. Table 5 also presents the prices of both the main products and by-products obtained in each case, offering a basis for a preliminary cost–benefit assessment.

**Table 5.** Comparison of oxidizing agent costs * and all product prices * for treating 1 metric ton of ZnS.

| Case | Chemical | Chemical Price $/t | Required/Obtained Amount, t/t ** | Total Cost/Benefit, USD/t |
|---|---|---|---|---|
| I | $(NH_4)_2S_2O_8$ | 680 | 6.549 | 4453 |
| II | $Fe_2(SO_4)_3$ | 370 | 11.48 | 4248 |
| III | $MnO_2$ | 535 | 3.327 | 1870 |
| | KI | 3000 | 0.030 | |
| I | $(NH_4)_2SO_4$ | 135 | 3.792 | 1715 |
| | Zn | 2713 | 0.4508 | |
| II | $FeSO_4 \times 7H_2O$ | 95 | 15.96 | 2649 |
| | Zn | 2713 | 0.4456 | |
| III | $MnSO_4 \times H_2O$ | 400 | 4.851 | 2885 |
| | Zn | 2713 | 0.4212 | |

* All chemical prices were obtained from Alibaba, while the zinc price was taken from the London Metal Exchange (LME) on 28 December 2024. ** Tones of input materials per tones of concentrate.

Based on the data in Table 5, both Case I and Case II are unprofitable, with total costs exceeding revenue by USD 2738 and USD 1599, respectively. In contrast, Case III, which employs our proposed leaching agents, indicates a profit of USD 1015, suggesting that this approach has promising commercial potential. While these figures provide an initial gauge of economic feasibility, further detailed assessments—potentially at pilot or industrial scales—are necessary to validate and refine these preliminary cost–benefit conclusions.

## 4. Conclusions

This work explored Zn leaching from sphalerite concentrates in sulfuric acid (40–80 °C, atmospheric pressure), employing $MnO_2$ alone and in combination with KI. Using $MnO_2$ by itself yielded low Zn extraction efficiencies, even at higher temperatures. SEM-EDS analysis revealed nonstoichiometric sulfur on marcasite grains and reduced oxygen content on pyrolusite grains, suggesting partial disulfide oxidation, elemental sulfur formation, and partial Mn(IV) reduction.

Temperature exerted a strong influence on Zn extraction, increasing leaching efficiency markedly from 40 °C to 80 °C due to faster reaction kinetics. At 80 °C, XRD confirmed the presence of crystalline sulfur phases, whereas no crystallinity was detected at 40 °C, implying that elevated temperatures promote sulfur crystallization. Despite potential passivation risks, higher temperatures ultimately provided net benefits in Zn recovery.

Introducing KI produced a nuanced relationship between oxidant availability and sulfur passivation. With 1 wt.% KI, Zn leaching reached 79.04%, but at 2 wt.% KI, increased sulfur passivation reduced the efficiency to 67.53%. Notably, 3 wt.% KI elevated the Zn recovery to 89.78% by oxidizing ZnS and transforming elemental sulfur into soluble compounds (e.g., $MnSO_4$), thereby reducing passivation.

Additionally, the kinetic analysis of the $MnO_2$ + KI system in sulfuric acid revealed a low activation energy of 27.65 kJ $mol^{-1}$. This low activation energy signifies that the leaching reactions proceed readily at relatively lower temperatures, making the process more energy-efficient and economically favorable. The enhanced reaction kinetics contribute to the overall effectiveness of the leaching system, facilitating higher Zn recovery rates.

Overall, this study confirms that Zn leaching from sphalerite is strongly governed by temperature and KI concentration, with 80 °C and 3 wt.% KI yielding the highest Zn recovery by mitigating sulfur passivation. Additionally, a preliminary cost–benefit analysis indicated that utilizing $MnO_2$ + KI can be more profitable than alternative oxidants, further supporting the practicality of this approach.

**Supplementary Materials:** The following supporting information can be downloaded at: https://www.mdpi.com/article/10.3390/met15010050/s1, Figure S1. Native silver/gold next to pyrite in the Zn concentrate sample; Figure S2. EDS analysis, spectrum 34 from Figure 10; Table S1. Chemical composition obtained from the spectrum 34 (marcasite); Figure S3. EDS analysis spectrum 35 from Figure 10; Figure S4. EDS analysis, spectrum 36 from Figure 10; Figure S5. EDS analysis, spectrum 46 from Figure 12; Figure S6. EDS analysis, spectrum 46 from Figure 12; Figure S7. EDS analysis, spectrum 48 from Figure 12; Figure S8. EDS analysis, spectrum 49 from Figure 12.

**Author Contributions:** Conceptualization, M.B., A.J. and M. S.; methodology, M.S. and B.M.; software, I.J., A.J. and D.A.; validation, M.B., M.S. and B.M.; formal analysis, M.B., A.J. and I.J.; investigation, D.A., M.B. and A.J.; resources, M.S. and B.M.; data curation, A.J., I.J., M.M. and D.A.; writing—original draft preparation, M.B. and A.J.; writing—review and editing, I.J., M.S., M.M. and B.M.; visualization, A.J., M.B. and M.M.; supervision, M.S.; project administration, M.S. and M.B. All authors have read and agreed to the published version of the manuscript.

**Funding:** This work was supported by the Ministry of Science and Technological Development and Innovation of the Republic of Serbia (Contract No. 451-03-66/2024-03/200023).

**Data Availability Statement:** The data presented in this study are available on request from the corresponding author. The data are not publicly available due to privacy.

**Conflicts of Interest:** The authors declare no conflicts of interest.

## References

1. Kania, H.; Saternus, M. Evaluation and Current State of Primary and Secondary Zinc Production—A Review. *Appl. Sci.* **2023**, *13*, 2003. [CrossRef]
2. Sokić, M.D.; Bugarčić, M.; Jovanović, A.A. Sphalerite Leaching in Acid Media: A Review. *Metall. Mater. Data* **2023**, *1*, 33–43. [CrossRef]
3. Aydogan, S. Dissolution Kinetics of Sphalerite with Hydrogen Peroxide in Sulphuric Acid Medium. *Chem. Eng. J.* **2006**, *123*, 65–70. [CrossRef]

4. Mahajan, V.; Misra, M.; Zhong, K.; Fuerstenau, M.C. Enhanced Leaching of Copper from Chalcopyrite in Hydrogen Peroxide–Glycol System. *Min. Eng.* **2007**, *20*, 670–674. [CrossRef]
5. Kartal, M.; Xia, F.; Ralph, D.; Rickard, W.D.A.; Renard, F.; Li, W. Enhancing Chalcopyrite Leaching by Tetrachloroethylene-Assisted Removal of Sulphur Passivation and the Mechanism of Jarosite Formation. *Hydrometallurgy* **2020**, *191*, 105192. [CrossRef]
6. Fan, Y.; Liu, Y.; Niu, L.; Jing, T.; Zhang, T. Separation and Purification of Elemental Sulfur from Sphalerite Concentrate Direct Leaching Residue by Liquid Paraffin. *Hydrometallurgy* **2019**, *186*, 162–169. [CrossRef]
7. Tian, L.; Zhang, T.A.; Liu, Y.; Lv, G.Z.; Tang, J.J. Oxidative Acid Leaching of Mechanically Activated Sphalerite. *Can. Metall. Q.* **2018**, *57*, 59–69. [CrossRef]
8. Al-Harahsheh, M.; Kingman, S. The Influence of Microwaves on the Leaching of Sphalerite in Ferric Chloride. *Chem. Eng. Process. Intensif.* **2007**, *46*, 883–888. [CrossRef]
9. Aydogan, S.; Aras, A.; Canbazoglu, M. Dissolution Kinetics of Sphalerite in Acidic Ferric Chloride Leaching. *Chem. Eng. J.* **2005**, *114*, 67–72. [CrossRef]
10. Lampinen, M.; Laari, A.; Turunen, I. Kinetic Model for Direct Leaching of Zinc Sulfide Concentrates at High Slurry and Solute Concentration. *Hydrometallurgy* **2015**, *153*, 160–169. [CrossRef]
11. Lorenzo-Tallafigo, J.; Iglesias-Gonzalez, N.; Romero, R.; Mazuelos, A.; Carranza, F. Ferric Leaching of the Sphalerite Contained in a Bulk Concentrate: Kinetic Study. *Min. Eng.* **2018**, *125*, 50–59. [CrossRef]
12. Nikkhou, F.; Xia, F.; Deditius, A.P. Variable Surface Passivation during Direct Leaching of Sphalerite by Ferric Sulfate, Ferric Chloride, and Ferric Nitrate in a Citrate Medium. *Hydrometallurgy* **2019**, *188*, 201–215. [CrossRef]
13. Nikkhou, F.; Xia, F.; Deditius, A.P.; Yao, X. Formation Mechanisms of Surface Passivating Phases and Their Impact on the Kinetics of Galena Leaching in Ferric Chloride, Ferric Perchlorate, and Ferric Nitrate Solutions. *Hydrometallurgy* **2020**, *197*, 105468. [CrossRef]
14. Bakhti, A.; Moghimi, H.; Bozorg, A.; Stankovic, S.; Manafi, Z.; Schippers, A. Comparison of Bioleaching of a Sulfidic Copper Ore (Chalcopyrite) in Column Percolators and in Stirred-Tank Bioreactors Including Microbial Community Analysis. *Chemosphere* **2024**, *349*, 140945. [CrossRef] [PubMed]
15. Babu, M.N.; Sahu, K.K.; Pandey, B.D. Zinc Recovery from Sphalerite Concentrate by Direct Oxidative Leaching with Ammonium, Sodium and Potassium Persulphates. *Hydrometallurgy* **2002**, *64*, 119–129. [CrossRef]
16. Sahu, S.K.; Sahu, K.K.; Pandey, B.D. Leaching of Zinc Sulfide Concentrate from the Ganesh-Himal Deposit of Nepal. *Metall. Mater. Trans. B* **2006**, *37*, 541–549. [CrossRef]
17. Zhukov, V.V.; Laari, A.; Lampinen, M.; Koiranen, T. A Mechanistic Kinetic Model for Direct Pressure Leaching of Iron Containing Sphalerite Concentrate. *Chem. Eng. Res. Des.* **2017**, *118*, 131–141. [CrossRef]
18. Duoqiang, L.; Jikun, W.; Yunhua, W.; Huaiwei, Z. Catalytic Role of Manganese in Autoclave Oxidation of Germanium-Rich Sphalerite Concentrates. *Can. J. Chem. Eng.* **2009**, *87*, 106–109. [CrossRef]
19. Li, C.; Wei, C.; Xu, H.S.; Li, M.; Li, X.; Deng, Z.; Fan, G. Oxidative Pressure Leaching of Sphalerite Concentrate with High Indium and Iron Content in Sulfuric Acid Medium. *Hydrometallurgy* **2010**, *102*, 91–94. [CrossRef]
20. Jovanović, G.; Bugarčić, M.; Petronijević, N.; Stopic, S.R.; Friedrich, B.; Marković, B.; Stanković, S.; Sokić, M. A Multifocal Study Investigation of Pyrolyzed Printed Circuit Board Leaching. *Metals* **2022**, *12*, 2021. [CrossRef]
21. Min, X.B.; Xie, X.D.; Chai, L.Y.; Liang, Y.J.; Li, M.; Ke, Y. Environmental Availability and Ecological Risk Assessment of Heavy Metals in Zinc Leaching Residue. *Trans. Nonferrous Met. Soc. China* **2013**, *23*, 208–218. [CrossRef]
22. Ismael, M.R.C.; Carvalho, J.M.R. Iron Recovery from Sulphate Leach Liquors in Zinc Hydrometallurgy. *Min. Eng.* **2003**, *16*, 31–39. [CrossRef]
23. Jorjani, E.; Ghahreman, A. Challenges with Elemental Sulfur Removal during the Leaching of Copper and Zinc Sulfides, and from the Residues; a Review. *Hydrometallurgy* **2017**, *171*, 333–343. [CrossRef]
24. Tian, L.; Gong, A.; Wu, X.G.; Xu, Z.F.; Zhang, T.A.; Liu, Y.; Wei, K.X.; Yu, Z.L. Oxygen Pressure Acid Leaching of Artificial Sphalerite Catalyzed by $Fe^{3+}/Fe^{2+}$ Self-Precipitation. *J. Cent. South Univ.* **2020**, *27*, 1703–1713. [CrossRef]
25. Parker, A.J.; Paul, R.L.; Power, G.P. Electrochemical Aspects of Leaching Copper from Chalcopyrite in Ferric and Cupric Salt Solutions. *Aust. J. Chem.* **1981**, *34*, 13–34. [CrossRef]
26. Toro, N.; Pérez, K.; Saldaña, M.; Jeldres, R.I.; Jeldres, M.; Cánovas, M. Dissolution of Pure Chalcopyrite with Manganese Nodules and Waste Water. *J. Mater. Res. Technol.* **2020**, *9*, 798–805. [CrossRef]
27. Li, L.; Soleymani, M.; Ghahreman, A. New Insights on the Role of Lattice-Substituted Silver in Catalytic Oxidation of Chalcopyrite. *Electrochim Acta* **2021**, *369*, 137652. [CrossRef]
28. Jha, M.K.; Kumar, V.; Singh, R.J. Review of Hydrometallurgical Recovery of Zinc from Industrial Wastes. *Resour. Conserv. Recycl.* **2001**, *33*, 1–22. [CrossRef]
29. Sharp, J.H.; Brindley, G.W.; Achar, B.N.N. Numerical Data for Some Commonly Used Solid State Reaction Equations. *J. Am. Ceram. Soc.* **1966**, *49*, 379–382. [CrossRef]

30. Sokić, M.D.; Marković, B.; Živković, D. Kinetics of Chalcopyrite Leaching by Sodium Nitrate in Sulphuric Acid. *Hydrometallurgy* **2009**, *95*, 273–279. [CrossRef]
31. Sokić, M.; Marković, B.; Matković, V.; Živković, D.; Štrbac, N.; Stojanović, J. Kinetics and Mechanism of Sphalerite Leaching by Sodium Nitrate in Sulphuric Acid Solution. *J. Min. Metall. Sect. B Metall.* **2012**, *48*, 185–195. [CrossRef]
32. Sokić, M.; Marković, B.; Stanković, S.; Kamberović, Ž.; Štrbac, N.; Manojlović, V.; Petronijević, N. Kinetics of Chalcopyrite Leaching by Hydrogen Peroxide in Sulfuric Acid. *Metals* **2019**, *9*, 1173. [CrossRef]
33. Ramette, R.W.; Sandford, R.W. Thermodynamics of Iodine Solubility and Triiodide Ion Formation in Water and in Deuterium Oxide. *J. Am. Chem. Soc.* **1955**, *77*, 1068. [CrossRef]
34. Winarko, R.; Dreisinger, D.B.; Miura, A.; Fukano, Y.; Liu, W. Iodine-Assisted Chalcopyrite Leaching in Ferric Sulfate Media: Kinetic Study under Fully Controlled Redox Potential and PH. *Hydrometallurgy* **2022**, *208*, 105797. [CrossRef]
35. Chanda, M.; Rempel, G.L. Catalysed Air Oxidation of Thiosulfate and Tetrathionate for Pollution Abatement. *Appl. Catal.* **1985**, *19*, 33–48. [CrossRef]
36. Druschel, G.K.; Hamers, R.J.; Banfield, J.F. Kinetics and Mechanism of Polythionate Oxidation to Sulfate at Low PH by $O_2$ and $Fe^{3+}$. *Geochim Cosmochim Acta* **2003**, *67*, 4457–4469. [CrossRef]
37. Qu, B.; Deng, L.; Deng, B.; He, K.; Liao, B.; Su, S. Oxidation Kinetics of Dithionate Compound in the Leaching Process of Manganese Dioxide with Manganese Dithionate. *React. Kinet. Mech. Catal.* **2018**, *123*, 743–755. [CrossRef]
38. Picazo-Rodríguez, N.G.; Soria-Aguilar, M.D.J.; Martínez-Luévanos, A.; Almaguer-Guzmán, I.; Chaidez-Félix, J.; Carrillo-Pedroza, F.R. Direct Acid Leaching of Sphalerite: An Approach Comparative and Kinetics Analysis. *Minerals* **2020**, *10*, 359. [CrossRef]
39. Weisener, C.G.; Smart, R.S.C.; Gerson, A.R. A Comparison of the Kinetics and Mechanism of Acid Leaching of Sphalerite Containing Low and High Concentrations of Iron. *Int. J. Min. Process.* **2004**, *74*, 239–249. [CrossRef]
40. Weisener, C.G.; Smart, R.S.C.; Gerson, A.R. Kinetics and Mechanisms of the Leaching of Low Fe Sphalerite. *Geochim Cosmochim Acta* **2003**, *67*, 823–830. [CrossRef]
41. Debernardi, G.; Carlesi, C. Chemical-Electrochemical Approaches to the Study Passivation of Chalcopyrite. *Miner. Process. Extr. Metall. Rev.* **2013**, *34*, 10–41. [CrossRef]
42. Hackl, R.P.; Dreisinger, D.B.; Peters, E.; King, J.A. Passivation of Chalcopyrite during Oxidative Leaching in Sulfate Media. *Hydrometallurgy* **1995**, *39*, 25–48. [CrossRef]
43. Peng, P.; Xie, H.; Lu, L. Leaching of a Sphalerite Concentrate with $H_2SO_4$–$HNO_3$ Solutions in the Presence of $C_2Cl_4$. *Hydrometallurgy* **2005**, *80*, 265–271. [CrossRef]
44. Karimi, S.; Rashchi, F.; Moghaddam, J. Parameters Optimization and Kinetics of Direct Atmospheric Leaching of Angouran Sphalerite. *Int. J. Min. Process.* **2017**, *162*, 58–68. [CrossRef]
45. Santos, S.M.C.; Machado, R.M.; Correia, M.J.N.; Reis, M.T.A.; Ismael, M.R.C.; Carvalho, J.M.R. Ferric Sulphate/Chloride Leaching of Zinc and Minor Elements from a Sphalerite Concentrate. *Min. Eng.* **2010**, *23*, 606–615. [CrossRef]

metals

MDPI

Article

# Innovative Integration of Citric Acid Leaching and Electrodialysis for Selective Lithium Recovery from NMC Cathode Material

**Soukayna Badre-Eddine [1,2], Laurence Muhr [2] and Alexandre Chagnes [1,*]**

[1] CNRS, GeoRessources, Université de Lorraine, F-54000 Nancy, France; soukayna.badre-eddine@univ-lorraine.fr
[2] CNRS, LRGP, Université de Lorraine, F-54000 Nancy, France; laurence.muhr@univ-lorraine.fr
* Correspondence: alexandre.chagnes@univ-lorraine.fr

**Abstract:** With the growing demand for metals driven by technological advancements and population growth, recycling lithium-ion batteries has become vital for protecting the environment and recovering valuable materials. Developing sustainable recycling technologies is now more essential than ever. This paper focuses on using electrodialysis to process a leach solution of $LiNi_{0.33}Mn_{0.33}Co_{0.33}O_2$ (NMC 111) cathode materials leached with citric acid. This study demonstrates that the complexing properties of citrate anions contribute to the efficient separation of Li from Ni, Co, and Mn by electrodialysis. This is achieved by promoting the formation of anionic species for Ni, Co, and Mn while maintaining Li in its cationic form. The leach solution was produced under the following optimal experimental conditions to reach a final pH of 5 and high leaching efficiency: a citric acid concentration of 1 mol $L^{-1}$, a leaching temperature of 45 °C, a leaching time of 5 h, a liquid/solid ratio of 100 g/L, and 8 vol.% $H_2O_2$. These conditions resulted in leaching efficiencies of 89.3% for Ni, 95.1% for Co, 77.1% for Mn, and 92.9% for Li. This solution led to the formation of a lithium-rich supernatant and a precipitate. The supernatant was then used as the feed solution for electrodialysis. Pure lithium was successfully separated with a faradic efficiency of 71.4% with a commercial cation-exchange membrane. This strategy enables selective lithium recovery while minimizing membrane fouling during the process.

**Keywords:** citric acid; electrodialysis; hydrometallurgy; leaching lithium-ion battery; NMC111; recycling

## 1. Introduction

Currently, the recycling of lithium-ion batteries (LIBs) primarily involves pyrometallurgical and hydrometallurgical processes [1]. While pyrometallurgy was initially the most common approach, it struggles to effectively separate metals, leading to metal losses, particularly lithium [2]. Hydrometallurgy, which relies on chemical solubilization in aqueous media, is often employed downstream to enable selective metal recovery through leaching, purification, and separation steps. Leaching consists of metal dissolution using acids, bases, or complexing agents, followed by the selective recovery and purification of leached metals through techniques like solvent extraction, ion exchange, precipitation, or crystallization [3]. At the end of the process, metals can be recovered as hydroxide, sulfate, chloride, or carbonate salts, which can be reused in the manufacturing chain (Figure 1a). However, lithium extraction typically occurs as the final step in the LIB recycling process, following the recovery of the other metals (Co, Ni, Fe, Mn, Al, and Cu). This is because leaching solutions usually have a pH between 0 and 2, while lithium precipitates at higher pH levels, i.e., pH > 13 for lithium hydroxide precipitation and starting at pH~4 with maximum efficiency above pH 8 for lithium carbonate precipitation [4]. Consequently, each extraction step can lead to lithium losses of up to 30% through adsorption or co-precipitation [4]. Nevertheless, combining solvent extraction steps can reduce these losses to around 15%.

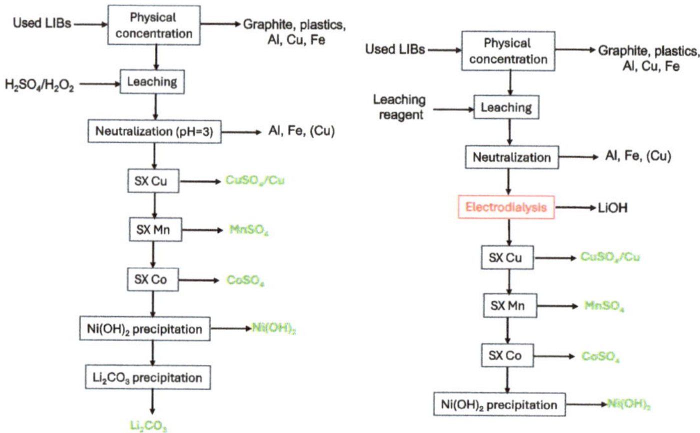


(**a**) (**b**)

**Figure 1.** (**a**) Typical hydrometallurgical flowsheet and (**b**) implementation of electrodialysis in battery recycling flowsheet (SX: solvent extraction).

Therefore, it is essential to develop an eco-friendly and sustainable technology for extracting metals from used lithium-ion batteries that allows obtaining lithium right from the beginning of the leachate treatment process, as illustrated in Figure 1b. Electrodialysis presents a promising solution for this initial step, offering high selectivity and purity in lithium recovery regardless of the presence of other metal ions. Its integration into hydrometallurgical processes is gaining attention, particularly in the lithium sector, whether for lithium production from spodumene or for metal recovery from used LIBs [5]. Two general approaches have been reported in the literature to separate lithium from cations in leach solutions produced by digesting cathode materials from spent LIBs. The first method, known as selective electrodialysis (SED), relies on the use of monovalent ion-exchange membranes to separate lithium from divalent cations such as Co, Ni, and Mn. Table 1 summarizes recent works focused on this technique for the selective separation of lithium from other metals present in LIB leachates. Gmar et al. [6] demonstrated that SED effectively separates and concentrates lithium from leach solutions of NMC111 cathode materials. Using sulfuric acid at pH 2.8, the process achieved a faradic efficiency of 67% for lithium(I) transfer at a current density of 10 mA $cm^{-2}$. Yet, prolonged operation led to membrane clogging. These phenomena were caused by cobalt(II), nickel(II) and/or manganese(II) precipitation inside the membranes due to the local increase in pH near the vicinity of the membrane under certain conditions. Similarly, Afifah et al. [7] employed SED to separate lithium from cobalt. They observed precipitation into the membrane at high voltage (15 V) and a decrease in current efficiency from 25% to 8% caused by a local increase in pH due to water dissociation. Membrane clogging was responsible for performance loss of the electrodialyzer and premature aging of the membranes.

**Table 1.** Lithium separation efficiency from cations in spent LIBs using electrodialysis (U: voltage; j: current density; $j_{Li}^+$ = Lithium flux; P(Li/M) = permselectivity; SEC(Li) = Specific Energy Consumption for lithium index).

| Methods | Cathode Materials + Complexing Agent | Membrane Type | U and/or j | Flowrate | Complementary Information | Ref. |
|---|---|---|---|---|---|---|
| Conventional electrodialysis and selective electrodialysis | Leach solution of NMC111 cathodic materials: 0.07 g $L^{-1}$ $Li^+$, 0.2 g $L^{-1}$ $Co^{2+}$, 0.2 g $L^{-1}$ $Ni^{2+}$, 0.18 g $L^{-1}$ $Mn^{2+}$, and $SO_4^{2-}$ + EDTA | Neosepta® CMS membrane, Neosepta® AMX, and Neosepta® CMX | 18 V (Stage 1) 18 V (Stage 2) 3 V (Stage 3) | 0.75 L $min^{-1}$ | $j_{Li}^+$ = 0.165 mol/($m^2$/h) $S_{Li^+/Mn^+}$ = 92% | [5] |
| Conventional electrodialysis | Leached solution produced from spent $LiCoO_2$ [$Li^+$] = [$Co^{2+}$] = 0.02 mol $L^{-1}$ + EDTA | 1 AEM (Selemion AMV), 1 CEM (Selemion CMV), and two bipolar membranes (Neosepta® BP-1E) | 20 V | 0.375 L $min^{-1}$ | R ($Li^+$) = R ($Co^{2+}$) = 99% | [8] |
| Selective electrodialysis | Typical mixing solution for used NMC [Li] = 2.6 g $L^{-1}$; [Co] = 7.88 g $L^{-1}$ [Mn] = 8.01 g $L^{-1}$; [Ni] = 4.4 g $L^{-1}$ + Sulfuric acid $H_2SO_4$ | 2 MEA 1 MEC monovalent (Neosepta®) | 12.5 mA $cm^{-2}$ | 100 mL $min^{-1}$ | $R_f$(Li) = 67.1% P(Li/Co) = 5.6 P(Li/Ni) = 6.1 P(Li/Mn) = 5.4 | [6] |
| Selective electrodialysis | Soluble nitrate salts ($LiNO_3$ + Co $(NO_3)_2 \cdot 6H_2O$) [Li] = 0.1 g $L^{-1}$; [Co] = 0.3 g $L^{-1}$ | 5 MEC (PC-MVK) 5 MEA (PC-MVA) | 15 V 15 A/$m^2$ | 15 L $h^{-1}$ | S (%) = 99.4% Lithium purity = 95.73% | [7] |
| Selective electrodialysis | [Li] = 0.027 g $L^{-1}$; [Co] = 0.108 g $L^{-1}$; [Ni]= 0.049 g $L^{-1}$ | PAN-5C8Q membrane | 5 V | 2.2 L.$h^{-1}$ | $j_{Li}^+$ =0.047 mol/($m^2$/h) | [9] |
| Selective electrodialysis | Leached solution produced from spent lithium ion [Li] = 1.616 g $L^{-1}$; [Co] = 0.059 g $L^{-1}$; [Mn] = 1.149 g $L^{-1}$ + Sulfuric acid $H_2SO_4$ | 1 MEC monovalent (Selemion CSO membrane) | 5 V | 17 $cm^3$/min | $Li_2CO_3$ purity = 99.6% SEC(Li) = 0.5 kW h/g $R_f$(Li) = 48.4% | [10] |
| Conventional electrodialysis | Typical mixing solution for used NMC [Li] = 3.27 g $L^{-1}$; [Co] = 0.25 g $L^{-1}$; [Mn] = 0.28 g $L^{-1}$; [Ni] = 0.25 g $L^{-1}$ + Sodium phosphate $Na_3PO_4$ | 1 MEC (Nafion 117) | 3.5 V | - | SC(P/Li) =3; $Li_2CO_3$ purity = 99.3%; SEC(Li) = 0.027 kWh/g; Current efficiency η(Li) = 50% | [11] |

The second method is conventional electrodialysis, which involves the use of standard ion exchange membranes for ion separation. However, its effectiveness in separating is limited by the lack of ion selectivity of standard membranes. In theory, lithium ions and other divalent cations cannot be separated by electrodialysis due to their similar charge densities.

Studies reported in the literature have focused on the use of electrodialysis in inorganic acids to separate lithium from divalent metals contained in aqueous solutions representative of leachates of LIB black masses (BMs). More and more works have aimed at investigating the use of inorganic acids to leach BMs or cathode materials from LIBs [12], whereas only a few papers have addressed the use of electrodialysis to process leaching solutions of BMs or cathode materials or representative solutions of leachates by organic acids. For instance, Chan et al. (2022) [5] studied the use of electrodialysis (ED) to separate lithium(I), cobalt(II), nickel(II), and manganese(II) contained in an aqueous solution resulting from the leaching of $LiNi_{1/3}Mn_{1/3}Co_{1/3}O_2$ cathode material by EDTA in the presence of $H_2O_2$. Nickel was recovered at approximately 99% at pH 2. In a second step, cobalt was separated with a yield of 87.3% at pH 3. Finally, the separation of lithium(I) and manganese(II) was performed using a monovalent-selective cation exchange membrane (Neosepta® CMS), with a separation yield of 99%. Likewise, Iizuka et al. [8] used bipolar membrane electrodialysis to separate lithium and cobalt contained in an aqueous solution produced by leaching $LiCoO_2$ electrodes with EDTA to form the complex $Co(EDTA)^{2-}$ at pH = 4. A faradic efficiency of 99% for both metals was achieved.

Among the organic acids, citric acid appears as a good reagent since it is considered an eco-friendly leaching agent due to its natural biodegradability, the absence of harmful gas emissions during the process, and its low cost, unlike conventional inorganic acids [13]. Citric acid was employed for its ability to effectively dissolve NMC electrodes and to form anionic complexes of cobalt(II), nickel(II), and manganese(II) without complexing lithium(I) within the pH range suitable for electrodialysis. Existing studies have consistently confirmed the effectiveness of citric acid as a lixiviant for the recovery of metals from LIB cathodes (LCO, LMO, and NCM). Table 2 presents a comparison of the leaching conditions of various types of cathodes by citric acid.

**Table 2.** Experimental conditions reported in the literature for cathode material leaching by citric acid.

| Cathode Materials | Citric Acid Concentration (mol $L^{-1}$) | Temperature (°C) | Leach Duration (min) | $H_2O_2$ Content (% in vol.) | S/L (g/L) | Leaching Yields (%) | pH of Leachate Solution | Ref. |
|---|---|---|---|---|---|---|---|---|
| $LiNi_{0.45}Mn_{0.4}Co_{0.15}O_2$ + $LiCoO_2$ | 1.5 | 95 | 30 | 2% | 20 | Li = 95.3%, Co = 89.8%, Ni = 93.6%, Mn = 94.4% | 2.5 | [14] |
| $LiCoO_2$ | 1 M | 80 °C | 60 | 4% | 20 | Li = 99%, Co = 99.21% | 1.29 | [15] |
| $LiNi_{0.33}Mn_{0.33}Co_{0.33}O_2$ (NMC111) | 1.5 M | 50 °C | 120 | 0% | 20 | Li = 82%, Co =63%, Ni = 74%, Mn = 95% | 2 | [16] |
| $LiNi_{0.33}Mn_{0.33}Co_{0.33}O_2$ (NMC111) | 2 M | 80 °C | 90 | 2% | 30 | Ni = 97% Co = 95%, Mn = 94% Li = 99% | 2.31 | [17] |
| $LiCoO_2$ (LCO) | 0.5 M | 60 °C | 300 | 6% | 25 | Li = 100%, Co = 96% | 2.5 | [18] |

Most of these works focused on parametric studies to find the optimal experimental conditions to leach LIB cathode materials or on comparisons with other organic acids, as investigated by Musariri B [19]. For the sake of illustration, Chen et al. [17] reported that the use of 1.5 mol $L^{-1}$ citric acid in the presence of 2% (vol.) $H_2O_2$ at 80 °C with a solid-to-liquid ratio of 30 g/L led to the dissolution of 99%, 97%, 95%, and 94% of Li, Ni, Co, and Mn, respectively, from an NMC111 cathode material.

Given the interesting properties of citric acid—not only its ability to leach NMC cathode materials but also its low toxicity and strong complexing capacity towards cobalt (II), nickel (II), and manganese (II)—it is expected to promote the efficient separation of lithium(I) from these divalent metals. Moreover, by increasing the solubility of cobalt (II), nickel (II), and manganese (II) through complexation, citric acid may help to prevent membrane fouling during downstream processing. Therefore, this study aims to investigate the combined use of citric acid leaching, electrodialysis, and crystallization to produce high-purity lithium salts for LIB production.

## 2. Materials and Methods

### 2.1. Materials

The NMC111 used in this study is a black powder primarily composed of Ni (19.175 wt%), Co (19.152 wt%), Mn (17.964 wt%), and Li (7.187 wt%), with trace amounts of Cu and Al (<0.003 wt%) and Fe (0.002 wt%). It was supplied by Xiamen Tob New Energy Technology Co., Ltd. (Xiamen, Fujian, China). Citric acid monohydrate ($C_6H_8O_7{\cdot}H_2O$, 99.9%, CAS: 5949-29-1) was obtained from Fisher Scientific (Paris, France) in its crystalline form, and hydrogen peroxide ($H_2O_2$, 30 wt% in water, CAS: 7722-84-1, 7732-18-5) was provided by Fisher Bioreagents (Paris, France). Lithium hydroxide (LiOH, 98%, CAS: 1310-65-2) was supplied by Sigma Aldrich (Paris, France). Monoelement standard solutions containing 1000 mg/L of Li, Ni, Co, or Mn in 2–5 wt% $HNO_3$ were used for the calibration of inductively coupled plasma-optical emission spectrometry (ICP-OES, ICAP 6000 series, Thermo Fisher Scientific, Waltham, MA, USA). Deionized water (>18.2 MΩ·cm) was used in all experiments. The filters used for solid–liquid separation after leaching experiments were Teflon membrane filters with a pore size of 0.45 μm and a capacity of 50 mL (SCP Science supplier, Paris, France). Standard buffer solutions at pH 2, 4.7, and 10 supplied by

VWR were used to calibrate the pH meter (VWR® pH 1100 L, Ref: 662-1657, Paris, France). The membranes employed for electrodialysis experiments were Neosepta® CMX-fg as cation-exchange membranes and Neosepta® AMX-fg as anion-exchange membranes. Both types of membranes from Astom Company (Tokyo, Japan) were supplied by Eurodia Industrie (Pertuis, France). The suffix "Fg" designates a commercial product type within the Neosepta membrane series. They are functionally equivalent to the classical CMX and AMX membranes commonly used in electrodialysis. The only difference lies in the reinforcing mesh. The classical CMX/AMX membranes use a PVC mesh, whereas CMX-Fg and AMX-Fg use a PE/PP (polyethylene/polypropylene) mesh.

### 2.2. Methods

#### 2.2.1. Analyses

Metal concentrations during leaching experiments were determined by microwave plasma-coupled atomic emission spectroscopy (4210 MP-AES, Agilent, Paris, France), whereas metal concentrations for electrodialysis experiments were determined using induced plasma-optical emission spectroscopy (ICP-OES, ICAP 6000 series, Thermo Fisher Scientific, Paris, France).

Total Organic Carbon (TOC) determination was performed with a Shimadzu TOC analyzer by measuring Non-Purgeable Organic Carbon (NPOC) in order to determine the citric acid concentration after leaching. A calibration curve was established using citric acid standard solutions with known carbon concentrations of 0.36 g/L, 0.50 g/L, 0.72 g/L, and 1.00 g/L (the citric acid standard solutions were prepared by dissolving appropriate amounts of citric acid monohydrate in deionized water). The resulting calibration equation was $y = 1723.8x$ with a correlation coefficient $R^2 = 0.9998$, where x represents the carbon concentration in g/L and y denotes for the signal area. The analyzed samples were subjected to a 100-fold dilution.

X-ray diffraction spectroscopy (XRD, Malvern Instruments, Malvern, UK) was performed to determine the crystallographic phases in the precipitates, and data were processed with Match! 4 software by using the Crystallography Open Database (COD: 4504918).

#### 2.2.2. Speciation Calculations

The speciation diagrams of lithium(I), nickel(II), manganese(II), and cobalt(II) in citric acid at different pH values were calculated using the HYDRA-MEDUSA software v1 (Make Equilibrium Diagrams Using Sophisticated Algorithms) [20], as the performance of metal separation by electrodialysis is highly dependent on complexation reactions and pH.

#### 2.2.3. Leaching

Leaching experiments were carried out at different solid-to-liquid (S/L) ratios. Preweighed cathode material samples were dissolved in a citric acid solution prepared by dissolving the required amount of citric acid in water. The experiments were conducted in 100 mL glass flasks. To avoid solid agglomeration at the bottom and ensure uniform mixing, the citric acid solution was added first, followed by the cathode material. Once mixed, hydrogen peroxide ($H_2O_2$) was added to start the leaching reaction. The flasks were placed in a thermostated mechanical shaker (shaker, Gerhardt France) set to 200 oscillations per minute and maintained at the desired temperature. At scheduled intervals, the flasks were removed from the shaker, and the mixtures were immediately filtered using Teflon membrane filters (pore size = 0.45 μm) to separate the liquid leachate from the undissolved solid residue. The pH of the liquid leachate was measured at room temperature using a calibrated pH meter. The resulting leachate was then analyzed to evaluate the leaching efficiency ($\%R(i)$):

$$\%R(i) = \frac{C_{measured}}{C_{total,calc}} \times 100\% \quad (1)$$

where $C_{measured}$ corresponds to the measured concentration and $C_{total,calc}$ represents the concentration that would be obtained after full leaching.

This methodology was applied to all samples, systematically varying the parameters to achieve the desired pH (citric acid concentration = 1 mol $L^{-1}$, S/L = 20–100 g/L, $H_2O_2$ content = 2–8% (vol.), and temperature = 30 or 45 °C). These ranges allowed for the evaluation of the influence of each parameter on leaching efficiency while optimizing the process to reach the target pH for electrodialysis separation.

For electrodialysis experiments, a 1 L leachate solution of NMC111 with citric acid was prepared in a thermostated double-jacketed leaching reactor (Heating Circulators, VWR® manufacturer, Paris, France) (Supplementary Materials, Figure S1). The diluted acid was added into the reactor under stirring at 600 rpm using an electronic rod stirrer (VOS 40 digital and VOS 60 control, VWR® manufacturer, Paris, France) followed by introducing the cathode material. The leaching solution was immediately filtered using vacuum filtration. The experimental conditions were selected based on the results of leaching studies.

#### 2.2.4. Electrodialysis

A four-compartment electrodialyzer was employed for the selective extraction of lithium(I) and divalent metals from an NMC111 leach solution. The compartments were separated by two cation-exchange membranes (CEMs) and an anion-exchange membrane (AEM) alternately positioned between platinized titanium electrodes (Figure 2). The area of each membrane was 40 $cm^2$. The system was operated in batch mode. For each compartment, an initial volume of 0.5 L of solution was introduced into a stirred beaker and pumped to ensure recirculation. This volume was selected to ensure complete filling of the compartment and tubes, to promote effective circulation, and to enable sampling throughout the 2 h experiment. The feed compartment was initially fed by the leach solution. The cathodic compartment (lithium recovery compartment) was initially fed with a 0.05 mol $L^{-1}$ LiOH solution. Lithium ions migrated into this compartment through a CEM. The metal recovery compartment, separated by an AEM, was filled with 0.5 mol $L^{-1}$ citric acid ($C_6H_8O_7$) for the recovery of metal–citric acid complexes. The electrolyte compartment, adjacent to the anode, contained initially 0.1 mol $L^{-1}$ $H_2SO_4$. This prevented the migration and possible oxidation of the metal complexes. The electrodialysis cell included four compartments.

A constant current of 4 A was applied to the electrodialyzer using DC power (DCS150-7E Sorensen, MB Electronique company, Buc France) for a duration of 2 h. Gases formed at the electrodes were extracted to ensure the safety of the installation. The flowrate of the peristaltic pumps (Masterflex™ L/S™ model 77250-62, Parmer instrument company, Vernon Hills, IL, USA) was set at 100 mL $min^{-1}$. Exactly 20 mL samples were collected every 30 min from the cathodic compartment (lithium recovery compartment) as well as the leachate and the metal recovery compartment for elemental analyses, pH measurements, and ionic conductivity measurements. These analyses enabled real-time monitoring of the separation efficiency between lithium and the transition metals present in the system.

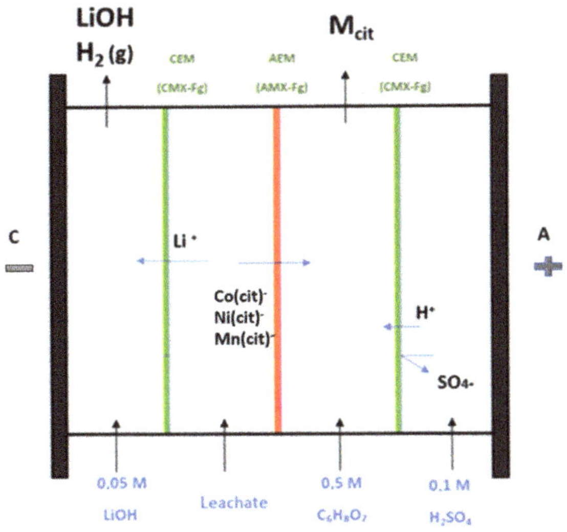


**Figure 2.** Scheme of the electrodialysis setup (CEM: cationic exchange membrane CMX-Fg; AEM: anionic exchange membrane AMX-Fg; $M_{cit}$: metal citrate species; A: anode; C: cathode; $Ni(cit)^-$, $Co(cit)^-$, and $Mn(cit)^-$: metal–citrate complexes of nickel, cobalt, and manganese, respectively).

The performance of the electrodialysis process was evaluated by calculating the faradic efficiency ($R_f$):

$$R_f(\%) = 100 \times \frac{zF\,\Delta n}{I\Delta t} \tag{2}$$

where $\Delta n$ (mol) is the number of moles of cations transported through the membrane, F is the Faraday constant (96,487 C $mol^{-1}$), z is the valence of the ion (for lithium, $z = 1$), I (A) is the applied current, and $\Delta t$ (s) is the electrodialysis operating time.

## 3. Results and Discussion

### 3.1. Speciation Calculations

Citric acid is a tricarboxylic acid with three carboxyl groups (-COOH) and one hydroxyl group (−OH) on a six-carbon chain ($C_6$) [21]. Its structure gives its acidic and complexing properties, with three dissociation constants at 25 °C ($pK_{a1}$ = 3.13, $pK_{a2}$ = 4.76, and $pK_{a3}$ = 6.39) [22]. In solution, it releases protons ($H^+$), facilitating the dissolution of metal oxides from the used cathodes ($LiCoO_2$, $LiNiO_2$, and $LiMnO_2$) by breaking the metal–oxygen bonds. The released metal ions ($Li^+$, $Co^{2+}$, $Ni^{2+}$, and $Mn^{2+}$) can then form bi- or tridentate complexes, as well as binuclear or multidentate structures with citric acid [23]. In the absence of an external reducing agent such as hydrogen peroxide ($H_2O_2$), citric acid itself can act as a mild reducing agent and undergo oxidation during the leaching process of the cathodes. For this, $H_2O_2$ is often added to promote the complete reduction of cobalt and manganese to their +II oxidation states, thus improving the leaching efficiency [24]. Although no specific data are available in literature regarding the leaching reaction of NMC111 by citric acid, an analogy can be made with those described for metal oxides of similar composition in the presence of $H_2O_2$. The overall reaction is represented by Equation (2), where HOA denotes citric acid and $OA^-$ its carboxylate anion, and citric acid is treated as a monophotonic acid (HOA) for simplification purposes [25]. However, in practice, citric acid is a triprotic acid and can form several partially deprotonated species ($H_2cit^-$, $Hcit^{2-}$, and $cit^{3-}$), which can lead to various metal–citrate complexes depending on the pH. Species such as $M(cit)^-$ and M(Hcit) may dominate the equilibrium, as discussed in the following section. Consequently, the actual speciation during the leaching mechanism is more complex than that depicted in the simplified Equation (3):

$$6\ LiNi_{1/3}Mn_{1/3}Co_{1/3}O_{2(s)} + 18\ HOA_{(aq)} + 3\ H_2O_{2(l)} \rightleftharpoons 2\ Ni(OA)_{2(aq)} + 2\ Mn\ (OA)_{2(aq)} + 2\ Co(OA)_{2(aq)} + 6\ LiOA_{(aq)} + 3\ O_{2(g)} + 12\ H_2O_{(l)} \quad (3)$$

According to the speciation diagram in Figure 3, Ni, Co, and Mn fully form anionic species with citric acid at a pH near to 5, while Li keeps its cationic form ($Li^+$) at this pH value.

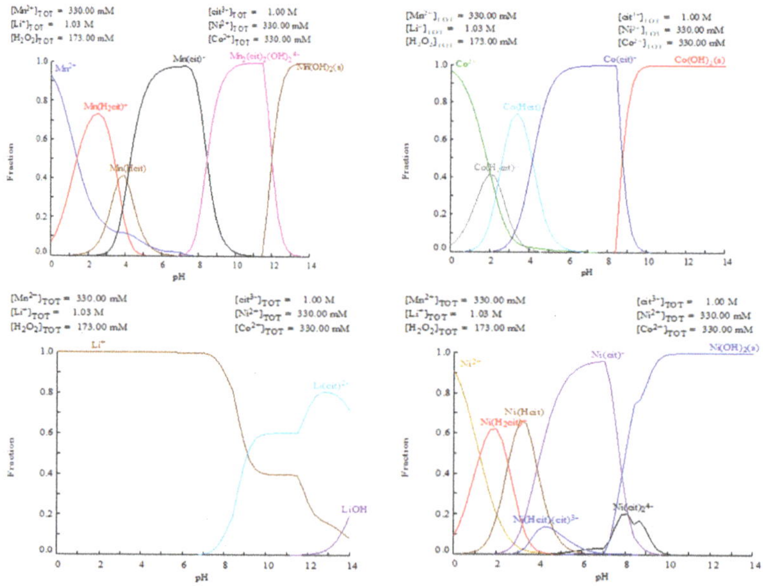

**Figure 3.** Speciation diagrams of manganese(II), cobalt(II), lithium(I), and nickel(II) in citric acid at 25 °C calculated for an aqueous solution, for which the composition is as follows: $[Mn^{2+}]$ = 330 mmol $L^{-1}$; $[Co^{2+}]$ = 330 mmol $L^{-1}$; $[Ni^{2+}]$ = 330 mmol $L^{-1}$; $[Li^+]$ = 1.03 mol $L^{-1}$; $[H_2O_2]$ = 173 mmol $L^{-1}$; and $[Cit^{3-}]$ = 1 mol $L^{-1}$ (calculation performed by using the Medusa software v1 with its database).

Therefore, pH=5 may be the optimal value for performing the efficient separation of Li(I) towards C(II), Ni(II), and Mn(II) by taking advantage of the difference in speciation. Furthermore, the complexation of the divalent metals may avoid their precipitation into the membranes and membrane clogging. This pH value can be reached by adjusting the pH after leaching with a base like sodium hydroxide (NaOH) or by choosing the appropriate value of the solid/liquid (S/L) ratio during leaching. This last option appears to be the best one since the use of NaOH for pH adjustment is responsible for the addition of sodium impurity in the leachate.

Therefore, the influence of the S/L ratio on pH and metal leaching efficiency was investigated using 1 mol $L^{-1}$ citric acid with 2% $H_2O_2$ at 30 °C (Figure 4a). Increasing the S/L ratio from 20 g/L to 100 g/L led to a pH rise from 2.4 to 3.6 after 24 h. Since electrodialysis should operate optimally at pH 4–5 according to speciation calculations (Figure 3), a further increase in the S/L ratio may be beneficial. However, excessive S/L ratios could lower the leaching efficiency due to acid depletion. After 24 h, leaching efficiencies for Mn, Ni, Co, and Li were 62.5%, 74.2%, 70.6%, and 81.6% (Figure 5a), respectively. From an industrial perspective, a higher S/L ratio improves scalability by enabling larger slurry processing.

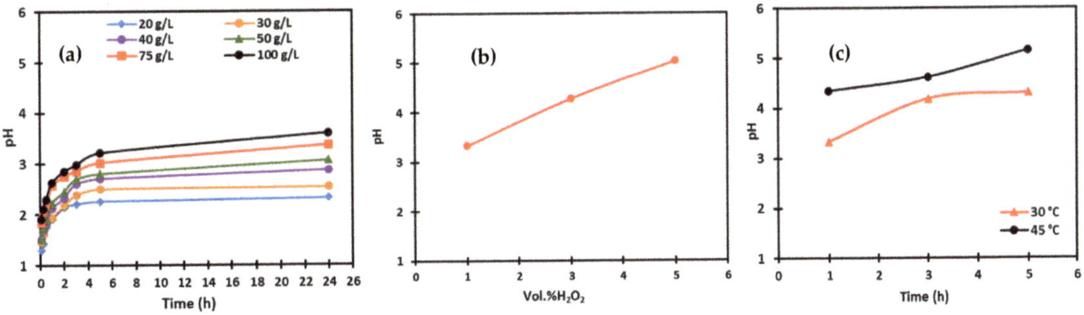


**Figure 4.** Effect of (**a**) solid-to-liquid ratio (S/L) (30 °C, %$H_2O_2$ (vol) = 2%), (**b**) $H_2O_2$ (%vol.) (S/L = 100 g/L, 30 °C, 24 h), and (**c**) temperature (S/L = 100 g/L, 8% $H_2O_2$) on the final pH value after leaching.

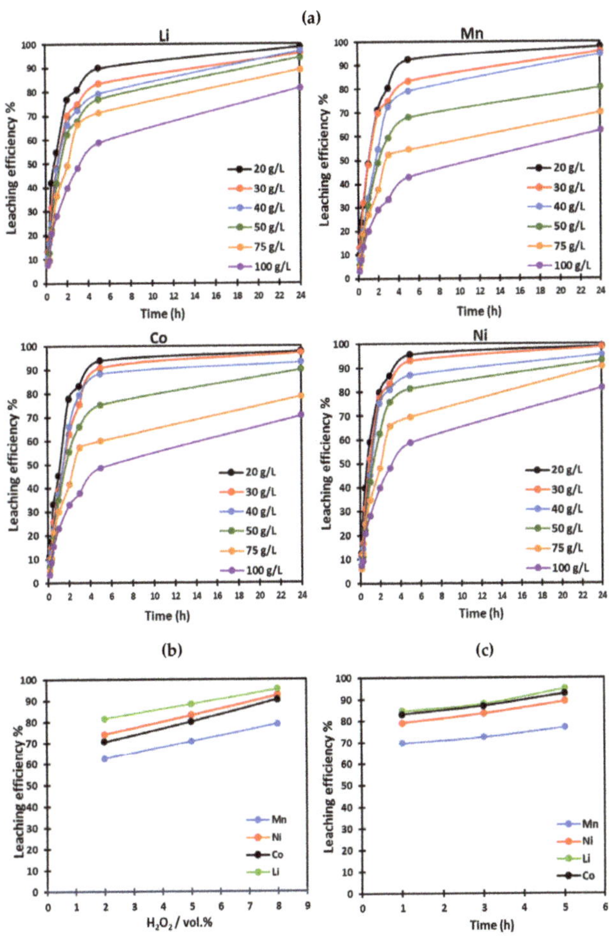


**Figure 5.** Effect on leaching efficiencies of (**a**) solid-to-liquid ratio (S/L) (30 °C, %$H_2O_2$(vol.) = 2%), (**b**) $H_2O_2$ (%vol.) (100 g/L, 30 °C, 24 h), and (**c**) time (100 g/L, 8% $H_2O_2$, and 45 °C).

The presence of hydrogen peroxide ($H_2O_2$) significantly affects pH values during the leaching process. Without $H_2O_2$, citric acid is consumed by oxidation, leading to a reduction of Co(III) to Co(II). In contrast, $H_2O_2$ prevents this redox reaction, preserving citric acid in the medium. However, the observed increase in pH with higher $H_2O_2$ concentrations suggests that citric acid may still be consumed through other pathways, such as complexation reactions.

As shown in Figure 4b, increasing the $H_2O_2$ concentration from 2% to 8% (vol.) at 30 °C and S/L=100 g/L raised the pH from 3.5 to 5 after 24 h. Leaching efficiencies for Mn, Ni, Co, and Li were 70.7%, 83.2%, 80.1%, and 88.35%, respectively (Figure 5b).

Temperature also has a significant impact on pH evolution during the leaching process. Increasing the temperature from 30 °C to 45 °C accelerated the pH rise from 4.3 to 5.1 over 5 h, indicating an acceleration of chemical reactions, as shown in Figure 4c. Leaching efficiencies of 77.1%, 89.3%, 95.1%, and 92.9% were reached for Mn, Ni, Co, and Li respectively (Figure 5c). This observation aligns with Yao et al. [26], who reported that leaching efficiency and pH follow similar trends over time, indicating that higher temperatures enhance both metal dissolution and pH increase as a result of acid consumption during leaching. Consequently, NMC111 leaching with 1 mol $L^{-1}$ citric acid in the presence of 8 vol. % $H_2O_2$ at 45 °C and S/L = 100 g/L for 5 h enables the production of a leach solution at pH 5, and it dissolves 89.3%, 95.1%, 77.1%, and 92.9% of Ni, Co, Mn, and Li, respectively.

### 3.2. *Crystallization After Leaching*

A pink precipitate was formed after storing the leachate at room temperature for four days, as shown in Figure S2 (Supplementary Materials). Around 81.5% manganese, 62.5% nickel, and 66.7% Co precipitated, while Li remained predominantly in the liquid phase since 83.8% of Li was found in the supernatant, reflecting its high solubility. This precipitation is advantageous for electrodialysis since it significantly reduces the presence of divalent metals in the leachate, and the solid formed may be used as a precursor for the synthesis of new cathode materials [25,27]. This precipitation step that precedes electrodialysis will be investigated to produce a lithium-rich supernatant and a precipitate that could be used as a potential PCAM material.

On the other hand, the citric acid concentration in the supernatant was determined using Total Organic Carbon (TOC) analysis. The carbon concentration in the leachate was determined to be m(C) = 43.6 g/L. Based on the molecular composition of citric acid and its molar weight, the final concentration was determined to be 0.605 mol $L^{-1}$. This indicates a loss of citric acid from the solution, since 1 mol $L^{-1}$ was initially used, most likely due to precipitation during the process. Nevertheless, this change in concentration does not significantly affect metal speciation (see Figure S3, Supplementary Materials). At pH 5, citric acid still forms stable anionic complexes with divalent metal ions, $Co^{2+}$, $Mn^{2+}$, and $Ni^{2+}$, while lithium remains in its cationic form.

The XRD pattern of the sample indicates a crystalline structure (black line in Figure 6) closely matching the reference pattern of tricobalt (II) citrate dihydrate, $C_{12}H_{38}Co_3O_{28}$ (COD 4504918, red line). This suggests that this compound is one of the major phases present in the precipitate. Mn and Ni are present in the precipitate but not detectable in the XRD pattern, likely due to the formation of amorphous citrate complexes or crystalline structures similar to cobalt citrate, resulting in overlapping diffraction peaks. Or the chemical similarity between $Co^{2+}$, $Mn^{2+}$, and $Ni^{2+}$ explains the possibility that the partial substitution of $Co^{2+}$ by $Mn^{2+}$ and $Ni^{2+}$ occurs within the same crystal lattice, without significantly altering the structure. Similar observations were reported in a previous study of Xuan et al. [25], where XRD analysis identified manganese citrate monohydrate ($Mn(HC_6H_5O_7){\cdot}H_2O$) in the precipitate after citric acid leaching despite different experimental conditions. This supports the hypothesis that various divalent metal citrates may co-precipitate, possibly forming mixed or structurally similar phases.

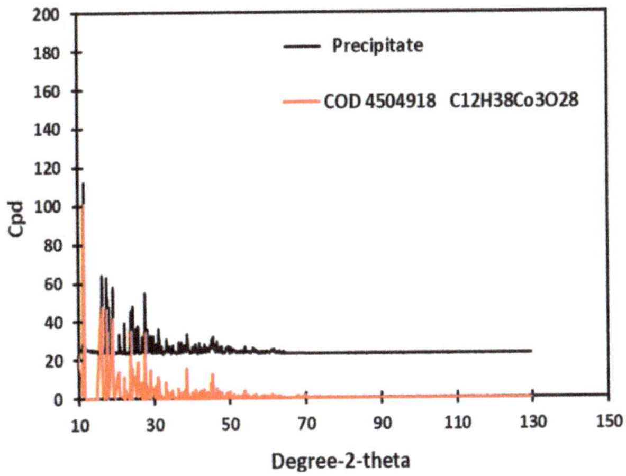


**Figure 6.** XRD pattern of the precipitate (black) and comparison with the XRD spectrum of tricobalt(II) citrate dihydrate ($C_{12}H_{38}Co_3O_{28}$).

### 3.3. Electrodialysis

Figure 6 shows the evolution of the lithium concentrations in the different compartments (lithium recovery, leachate, and metal recovery) during electrodialysis. A continuous and regular increase in the lithium concentration in the lithium recovery compartment was observed, from 0.05 mol $L^{-1}$ to approximately 0.49 mol $L^{-1}$ after 2 h, accompanied by a simultaneous decrease in the leachate compartment from 1 mol $L^{-1}$ to 0.69 mol $L^{-1}$. This evolution reflects an efficient and oriented transfer of lithium ions through the CEM. Simultaneously, the lithium concentration in the metal recovery compartment remains zero throughout the experiment, confirming the total absence of migration to this compartment. This behavior highlights the non-complexation of lithium by citric acid, thus facilitating its exclusive transport through the CEM. Conversely, transition metals (Ni, Co, and Mn), strongly complexed by citric acid, do not cross the CEM but migrate through the AEM into the metal recovery compartment. This is evidenced by the gradual color change observed in the solution, reflecting the progressive migration of these metal–ligand complexes. All of these results underline the selectivity of the process and its effectiveness for the targeted recovery of lithium in a highly pure form from multi-component leachates, with no clogging observed during operation (Supplementary Materials, Figure S4).

In order to calculate the number of moles of Li transferred to the recovery compartment over time, the mass balance must take into account changes in the volume of the various compartments due to sampling, water transfer by osmosis and electroosmosis, and slight evaporation. A notable decrease in leachate volume is observed in the corresponding compartment, leading to an asymmetry in the concentration curves of the two compartments (leachate and cathodic), as shown in Figure 7.

The calculation of the cumulative number of moles of Li transferred as a function of time revealed a linear increase over the first 90 min, with a transfer rate of $3.3 \times 10^{-5}$ mol $s^{-1}$. For a current of 4 A, this corresponds to a faradic efficiency of 78.7%. The transfer rate then slowed down between 90 and 120 min due to the decreasing concentration in the leachate compartment, which caused the imposed current to approach the limiting current. After 90 min, the value of the imposed current should have been reduced. The overall faradic efficiency for lithium recovery of the experiment was thus penalized, dropping to 71.4%. By reducing the current, much higher faradic efficiencies could be expected. To further recover lithium, the current should be reduced as the lithium concentration in the leachate compartment decreases.

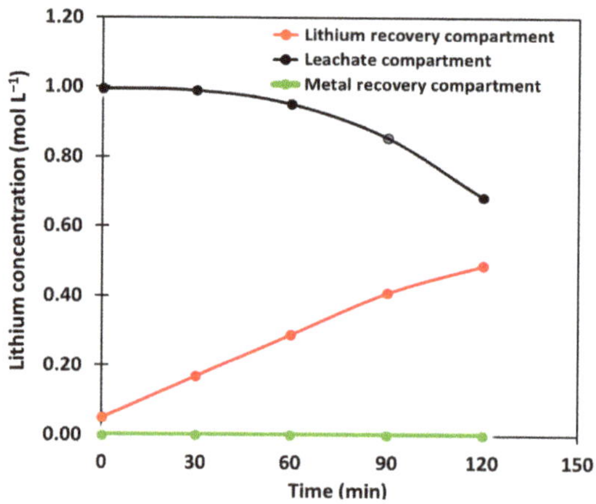


**Figure 7.** Lithium concentration in the different compartments during electrodialysis (I = 4A, flowrate = 100 mL $min^{-1}$, t = 2 h).

Table 3 shows the variation in pH and ionic conductivity as a function of time during electrodialysis in the leachate compartment, the lithium recovery compartment, and the metals recovery compartment.

**Table 3.** pH and ionic conductivity values recorded in each compartment as a function of time during electrodialysis (κ: ionic conductivity).

| | ED Compartments | | | | | |
|---|---|---|---|---|---|---|
| | $Li^+$ Recovery Compartment | | Leachate Compartment | | Metal Recovery Compartment | |
| Time (min) | pH | κ (mS $cm^{-1}$) | pH | κ (mS $cm^{-1}$) | pH | κ (mS $cm^{-1}$) |
| 0 | 12.75 | 9.26 | 4.8 | 23.88 | 1.60 | 5.42 |
| 30 | 13.47 | 31.64 | 4.85 | 23.53 | 1.66 | 5.48 |
| 60 | 13.55 | 51.60 | 4.85 | 23.74 | 1.98 | 5.55 |
| 90 | 13.59 | 70.70 | 4.75 | 21.62 | 2.04 | 5.62 |
| 120 | 13.67 | 91.00 | 4.56 | 18.90 | 2.08 | 5.88 |

Examination of this table shows an increase in pH from 12.75 to 13.67 in the lithium recovery compartment due to the formation of $OH^-$ ions by water reduction at the cathode, as well as an increase in ionic conductivity from 9.26 mS $cm^{-1}$ to 91 mS $cm^{-1}$ due to $Li^+$ migration and $OH^-$ formation.

In the leachate compartment, the pH remained relatively stable, changing slightly from 4.80 to 4.56. This pH stability can be attributed to the buffering effect of the medium, whose pH is close to that of $pKa_2$ of citric acid ($pK_{a2}$ = 4.76). It enables the limitation of pH fluctuations. This situation preserves the predominance of the $M(Cit)^-$ species, contributing to the stability of the targeted speciation throughout the process.

## 4. Conclusions

This study demonstrated the viability of conventional electrodialysis as an effective method for the highly selective separation of lithium from a citric acid-based leachate of

NMC111. The leaching process was found to promote the formation of anionic complexes of Ni(II), Co(II), and Mn(II) in the form of metal citrates, while lithium remained in its cationic form ($Li^+$). This speciation contrast led to the precipitation of transition-metal-rich solids and a lithium-enriched supernatant. This difference in speciation enabled the full potential of ion-exchange membranes to be exploited: anionic metal complexes migrate through anionic membranes, while lithium cations migrate through cationic membranes. The electrodialysis experiment resulted in the recovery of high-purity lithium with a faradic yield of up to 71.4%.

This study paves the way for an alternative approach combining leaching in organic media, electrodialysis, and crystallization, as depicted in Figure 8. After leaching the black mass in citric acid under the optimized conditions established in this study, a crystallization step precedes electrodialysis to maximize the removal of divalent cations—mainly manganese, cobalt, and nickel citrates—and produces a lithium-enriched solution compatible with electrodialysis. The stoichiometry can be adjusted with additional metal citrates before calcination to produce a pCAM suitable for the synthesis of new cathode materials for LIBS (CAM). A final electrodialysis step is then performed to obtain high-purity lithium hydroxide (LiOH). Recycling of the depleted leachate to the leaching step is proposed to improve process sustainability. Its impact on leaching efficiency remains to be explored in future work.

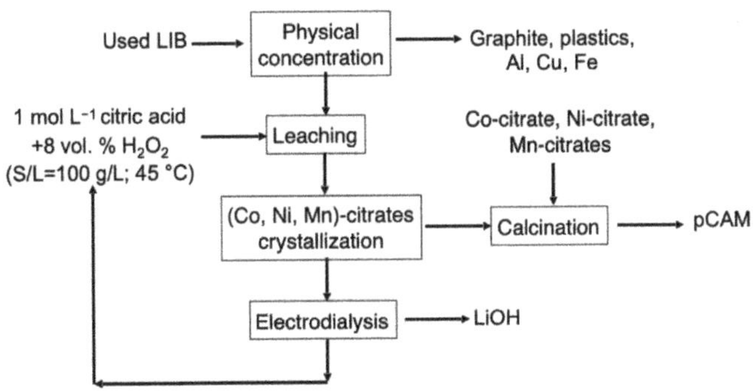


**Figure 8.** Alternative route for BM treatment by combining leaching, electrodialysis, and crystallization.

**Supplementary Materials:** The following supporting information can be downloaded at: https://www.mdpi.com/article/10.3390/met15060598/s1, Figure S1: Setup for leaching experiments using a thermostatic glass reactor. Figure S2: Precipitation observed in the leachate after 4 days of storage and Precipitate after filtration, washing and drying at 40 °C. Figure S3: Speciation diagrams of Mn(II), Co(II), Li(I) and Ni(II) in citric acid at 25° calculated for an aqueous solution which the composition is: $[Mn^{2+}]$ = 70 mmol $L^{-1}$; $[Co^{2+}]$ = 150 mmol $L^{-1}$; $[Ni^{2+}]$ = 170 mmol $L^{-1}$; $[Li^+]$ = 850 mol $L^{-1}$; $[H_2O_2]$ = 173 mmol $L^{-1}$; $[Cit^{3-}]$ = 0.6 mol $L^{-1}$ (calculation performed by using the Medusa software with its database). Figure S4: Electrodialysis experiments.

**Author Contributions:** Conceptualization, A.C.; methodology, A.C., L.M., and S.B.-E.; validation, A.C. and L.M.; formal analysis, A.C., L.M., and S.B.-E.; investigation, A.C., L.M., and S.B.-E.; resources, A.C.; writing—original draft, preparation, S.B.-E., A.C., and L.M.; writing—review and editing, A.C. and L.M.; visualization, S.B.-E.; supervision, A.C. and L.M.; project administration, A.C. and L.M.; funding acquisition, A.C. All authors have read and agreed to the published version of the manuscript.

**Funding:** This project was financially supported by the French government through the France 2030 investment plan managed by the French National Research Agency (ANR), «LULABAT» Project (ANR-22-PERE-0007), Reference number ANR-22-PERE-0007.

**Data Availability Statement:** The raw data supporting the conclusions of this article will be made available by the authors on request.

**Acknowledgments:** The authors gratefully acknowledge the SAMPL platform for performing the elemental analyses at the LRGP laboratory.

**Conflicts of Interest:** The authors declare no conflict of interest.

## References

1. Xu, J.; Thomas, H.R.; Francis, R.W.; Lum, K.R.; Wang, J.; Liang, B. A review of processes and technologies for the recycling of lithium-ion secondary batteries. *J. Power Sources* **2008**, *177*, 512–527. [CrossRef]
2. Brückner, L.; Frank, J.; Elwert, T. Industrial Recycling of Lithium-Ion Batteries—A Critical Review of Metallurgical Process Routes. *Metals* **2020**, *10*, 1107. [CrossRef]
3. Chagnes, A. *Métallurgie Extractive-Hydrométallurgie*; Techniques de l'ingénieur: Saint-Denis, France, 2022. (In French)
4. Marcinov, V.; Klimko, J.; Takáčová, Z.; Pirošková, J.; Miškufová, A.; Sommerfeld, M.; Dertmann, C.; Friedrich, B.; Oráč, D. Lithium Production and Recovery Methods: Overview of Lithium Losses. *Metals* **2023**, *13*, 1213. [CrossRef]
5. Chan, K.H.; Malik, M.; Azimi, G. Separation of lithium, nickel, manganese, and cobalt from waste lithium-ion batteries using electrodialysis. *Resour. Conserv. Recycl.* **2022**, *178*, 106076. [CrossRef]
6. Gmar, S.; Chagnes, A.; Lutin, F.; Muhr, L. Application of Electrodialysis for the Selective Lithium Extraction Towards Cobalt, Nickel and Manganese from Leach Solutions Containing High Divalent Cations/Li Ratio. *Recycling* **2022**, *7*, 14. [CrossRef]
7. Afifah, D.N.; Ariyanto, T.; Supranto, S.; Prasetyo, I. Separation of Lithium Ion from Lithium-Cobalt Mixture using Electrodialysis Monovalent Membrane. *Eng. J.* **2018**, *22*, 165–179. [CrossRef]
8. Iizuka, A.; Yamashita, Y.; Nagasawa, H.; Yamasaki, A.; Yanagisawa, Y. Separation of lithium and cobalt from waste lithium-ion batteries via bipolar membrane electrodialysis coupled with chelation. *Sep. Purif. Technol.* **2013**, *113*, 33–41. [CrossRef]
9. Siekierka, A.; Yalcinkaya, F. Selective cobalt-exchange membranes for electrodialysis dedicated for cobalt recovery from lithium, cobalt and nickel solutions. *Sep. Purif. Technol.* **2022**, *299*, 121695. [CrossRef]
10. Shi, M.; Diaz, L.A.; Klaehn, J.R.; Wilson, A.D.; Lister, T.E. $Li_2$ $CO_3$ Recovery through a Carbon-Negative Electrodialysis of Lithium-Ion Battery Leachates. *ACS Sustain. Chem. Eng.* **2022**, *10*, 11773–11781. [CrossRef]
11. Song, Y.; Zhao, Z. Recovery of lithium from spent lithium-ion batteries using precipitation and electrodialysis techniques. *Sep. Purif. Technol.* **2018**, *206*, 335–342. [CrossRef]
12. Villen-Guzman, M.; Arhoun, B.; Vereda-Alonso, C.; Gomez-Lahoz, C.; Rodriguez-Maroto, J.M.; Paz-Garcia, J.M. Electrodialytic processes in solid matrices. New insights into battery recycling. A review. *J. Chem. Technol. Biotechnol.* **2019**, *94*, 1727–1738. [CrossRef]
13. Gerold, E.; Schinnerl, C.; Antrekowitsch, H. Critical Evaluation of the Potential of Organic Acids for the Environmentally Friendly Recycling of Spent Lithium-Ion Batteries. *Recycling* **2022**, *7*, 4. [CrossRef]
14. Punt, T.; Van Wyk, A.P.; Bradshaw, S.M.; Akdogan, G. Citric Acid Leaching Performance at High Solid-to-Liquid Ratios for Lithium-Ion Battery Recycling. *Min. Metall. Explor.* **2024**, *41*, 3463–3474. [CrossRef]
15. Richard, M. *Recyclage Assisté des Métaux Stratégiques de Batteries Lithium-Ion Par Chauffage Micro-Ondes*; HAL Theses: Bengaluru, India, 2022. (In French)
16. Xuan, W. *Développement d'un Procédé Hydrométallurgique Pour le Recyclage des Électrodes Positives de Type NMC Contenues dans les Batteries Lithium-Ion Usagées*; HAL Theses: Bengaluru, India, 2022. (In French)
17. Chen, X.; Zhou, T. Hydrometallurgical process for the recovery of metal values from spent lithium-ion batteries in citric acid media. *Waste Manag. Res. J. Sustain. Circ. Econ.* **2014**, *32*, 1083–1093. [CrossRef]
18. Li, L.; Zhai, L.; Zhang, X.; Lu, J.; Chen, R.; Wu, F.; Amine, K. Recovery of valuable metals from spent lithium-ion batteries by ultrasonic-assisted leaching process. *J. Power Sources* **2014**, *262*, 380–385. [CrossRef]
19. Musariri, B.; Akdogan, G.; Dorfling, C.; Bradshaw, S. Evaluating organic acids as alternative leaching reagents for metal recovery from lithium ion batteries. *Miner. Eng.* **2019**, *137*, 108–117. [CrossRef]
20. Puigdomenech, I. *MEDUSA (Make Equilibrium Diagrams Using Sophisticated Algorithms)*; School of Chemical Science and Engineering, Royal Institute of Technology: Stockholm, Sweden, 2013. Available online: https://www.kth.se/che/medusa/downloads-1.386254 (accessed on 24 May 2025).
21. Loewenstein, A.; Roberts, J.D. The Ionization of Citric Acid Studied by the Nuclear Magnetic Resonance Technique. *J. Am. Chem. Soc.* **1960**, *82*, 2705–2710. [CrossRef]
22. Gergoric, M.; Ravaux, C.; Steenari, B.-M.; Espegren, F.; Retegan, T. Leaching and Recovery of Rare-Earth Elements from Neodymium Magnet Waste Using Organic Acids. *Metals* **2018**, *8*, 721. [CrossRef]
23. Francis, A.; Dodge, C. *Bio/PhotoChemical Separation and Recovery of Uranium*; Office of Scientific and Technical Information: Oak Ridge, TN, USA, 2008.

24. Li, L.; Ge, J.; Wu, F.; Chen, R.; Chen, S.; Wu, B. Recovery of cobalt and lithium from spent lithium ion batteries using organic citric acid as leachant. *J. Hazard. Mater.* **2010**, *176*, 288–293. [CrossRef]
25. Xuan, W.; Chagnes, A.; Xiao, X.; Olsson, R.T.; Forsberg, K. Antisolvent Precipitation for Metal Recovery from Citric Acid Solution in Recycling of NMC Cathode Materials. *Metals* **2022**, *12*, 607. [CrossRef]
26. Yao, L.; Feng, Y.; Xi, G. A new method for the synthesis of $LiNi_{1/3}Co_{1/3}Mn_{1/3}O_2$ from waste lithium ion batteries. *RSC Adv.* **2015**, *5*, 44107–44114. [CrossRef]
27. Chagnes, A.; Forsberg, K. Raw Material Supply for Lithium-Ion Batteries in the Circular Economy. *Metals* **2023**, *13*, 1590. [CrossRef]

MDPI AG
Grosspeteranlage 5
4052 Basel
Switzerland
Tel.: +41 61 683 77 34

*Metals* Editorial Office
E-mail: metals@mdpi.com
www.mdpi.com/journal/metals

www.ingramcontent.com/pod-product-compliance
Lightning Source LLC
Chambersburg PA
CBHW061618210926
51850CB00009B/2367